ARCHITECTUR
36 Bedford S
Telep

THIS BOOK IS TO
THE LAST

24 APR 2009

Architectural Association

54041000594670

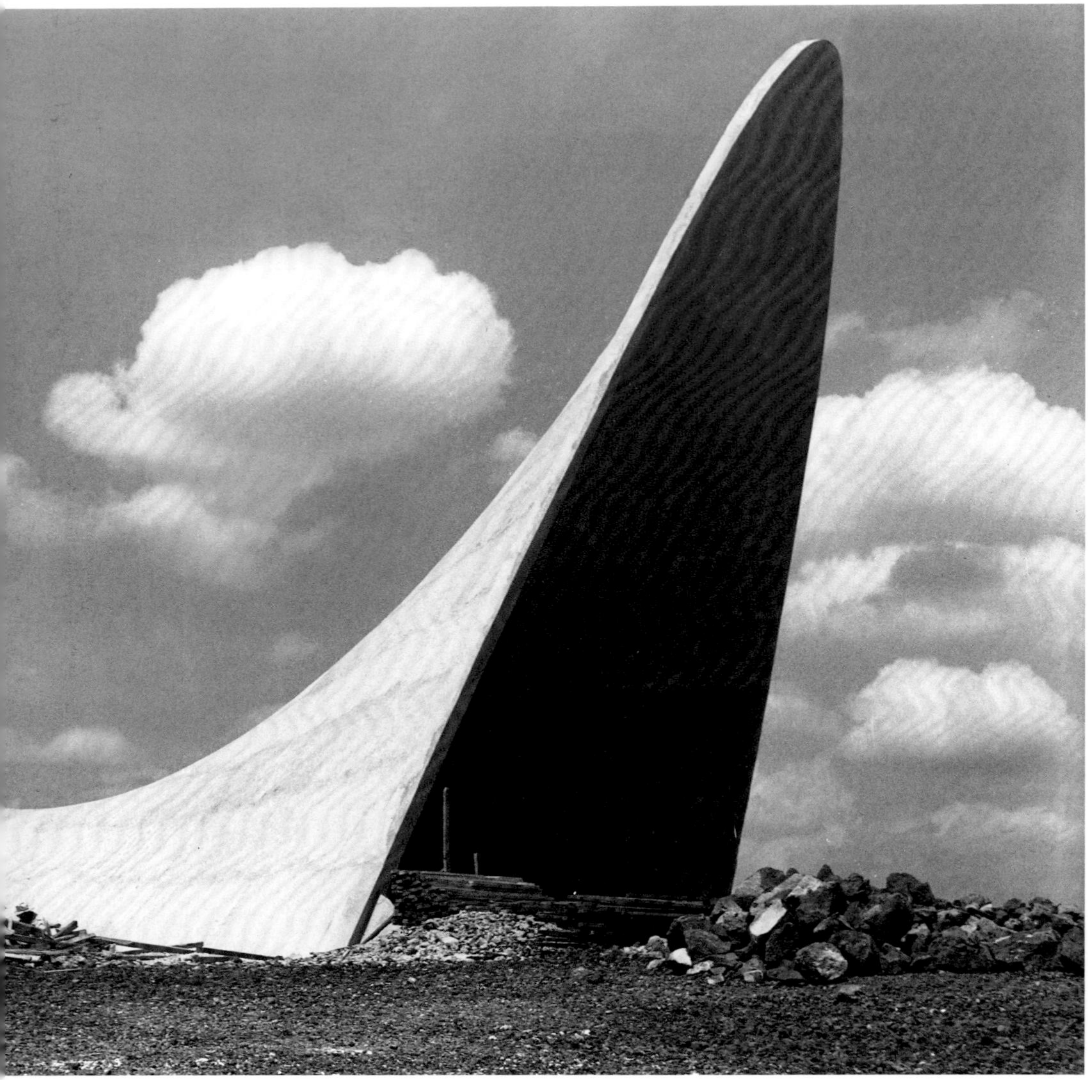

Seven Structural Engineers: The Felix Candela Lectures

Edited by
Guy Nordenson

THE MUSEUM OF MODERN ART, NEW YORK

The annual Felix Candela Lectures were given at The Museum of Modern Art, New York, Princeton University, and MIT, from 1998 to 2005 in honor of the great builder and designer Felix Candela.

This publication was made possible by Loren Pack and Rob Beyer.

Produced by the Department of Publications,
The Museum of Modern Art, New York
Edited by Joanne Greenspun and David Frankel
Designed by Antony Drobinski, Emsworth Design, Inc.
Production by Elisa Frohlich
Printed and bound by Editoriale Bortolazzi-Stei, S.R.L., Verona, Italy

Published by The Museum of Modern Art
11 West 53 Street, New York, New York 10019-5497

© 2008 by The Museum of Modern Art, New York

Copyright credits for certain illustrations are cited on page 177.
All rights reserved.

Distributed in the United States and Canada by
D. A. P./ Distributed Art Publishers, New York
Distributed outside the United States and Canada by
Thames & Hudson Ltd., London

Library of Congress Control Number: 2007921200
ISBN: 978-0-87070-703-2

Front cover: (above), Felix Candela. Chapel, Lomas de Cuernavaca (seen under construction); (below), Eladio Dieste. Massaro Agroindustries, Montevideo (detail); Cecil Balmond. CCTV Headquarters, Beijing; Heinz Isler. Wyss Garden Center, Solothurn, Switzerland (detail); Mamoru Kawaguchi. Fabric Carp, Kazo, Japan; Christian Menn, Ganter Bridge, Simplon Pass, Switzerland; Jörg Schlaich. DG Bank Building, Berlin; Leslie E. Robertson. Twin Towers, World Trade Center, New York

Back cover: Jörg Schlaich. Zaragoza Stadium, Zaragoza, Spain (detail)

Frontispiece: Felix Candela. Chapel, Lomas de Cuernevaca, partial view of structure under construction

Printed in Italy

Contents

Preface   7
TERENCE RILEY

Constellations   8
GUY NORDENSON

The Felix Candela Lectures

Eladio Dieste: A Principled Builder   30
STANFORD ANDERSON

Informal Networks   48
CECIL BALMOND

A Life in Structural Engineering   66
LESLIE E. ROBERTSON

Shell Structures: Candela in America and   86
What We Did in Europe
HEINZ ISLER

The Design of Structures—from Hard to Soft   102
MAMORU KAWAGUCHI

The Art of Bridge Design   122
CHRISTIAN MENN

On the Cultural and Social Responsibility   140
of the Structural Engineer
JÖRG SCHLAICH

Thin-Shell Concrete Structures: The Master Builders   160
DAVID BILLINGTON AND MARIA M. GARLOCK

Authors' Biographies   175
Acknowledgments   178
Photograph Credits   181
Index   183
Trustees of The Museum of Modern Art   188

Felix Candela with paper model of hyperbolic
paraboloid shell, Mexico, c. 1956

Preface

The Museum of Modern Art's first exhibition of the work of an engineer—*Robert Maillart: Engineer*—held in 1947, introduced a wide audience to Maillart's elegantly efficient Alpine structures and also to the idea that engineering, at its very best, could be presented at an institution dedicated to modern culture in all its forms. The Felix Candela Lectures, which commenced in 1998, reintroduced the subject of engineering—both in terms of its technical and its domestic dimensions—as a critical part of the Museum's design programs.

Felix Candela was born in Spain in 1910 and immigrated to Mexico after Franco's takeover of the government in 1939. Over the course of his career, Candela designed dozens of notable structures, the most innovative of which were characterized by daringly thin shells of reinforced concrete. In projects such as his 1960 Bacardi Rum Factory in Mexico City, Candela's designs called for lightweight, intersecting vaults that created a clear-span structure that maximized the flexibility of the floor plan. Transcending its utilitarian conception, his design also endowed the structure with spatial and formal characteristics—refined as sculpture—that are typical of his best works.

I would like to thank Guy Nordenson, Professor of Structural Engineering and Architecture at Princeton University, for his role in helping to conceive and organize the Candela lecture series. In his teaching, writing, and in his own engineering practice, Guy has consistently pursued the art of engineering in the sense that Candela understood it.

I would also like to note here what a great pleasure it was, when I had just graduated from architecture school, to work with Candela for a time. It was then that I also met his wife, Dorothy, whose support and encouragement, for which I am very grateful, has been a critical element in the successful realization of the Candela lecture series.

Terence Riley
former Philip Johnson Chief Curator of the Department of Architecture and Design, The Museum of Modern Art, New York

❶ Felix Candela. Cosmic Rays Pavilion, National University of Mexico, Mexico City

❷ Felix Candela. Signpost, Lake Tequesquitengo, Morelos, Mexico

Constellations
GUY NORDENSON

Ideas and Things

In Candela's work we have then an example of how complete mastery by one mind of all the facts affecting a design can produce that balanced perfection which makes a building or structure into a work of art.[1]—Ove Arup

In his book of essays, *El Arco y la lira (The Bow and the Lyre)*, the Mexican diplomat and poet Octavio Paz (1914–1998) writes that "justice and order are categories of being," that "both political and cosmic justice are not properly laws that are over the nature of things, but things themselves in their mutual movement, engendering themselves and devouring each other, produce justice."[2] Felix Candela, born January 27, 1910, embraced a strong sense of justice and order in his life and work. In his early years, growing up, studying and practicing architecture in Madrid, Candela led an active life of sports (he was Spain's national ski champion in 1932), mountain climbing, engineering, and design, and, from 1936, absorbed "some of the lessons portioned out by revolution and civil war."[3] At the end of the Spanish Civil War, he was lucky to be selected among the few hundred of more than 50,000 Republicans imprisoned in the French concentration camps near Perpignan who were evacuated to Mexico on a ship chartered by the Society of Friends. He arrived there in June 1939, working first in Acapulco, then in Mexico City, and in ten years was able to establish himself as a builder and designer. There followed in the ten years from the Fernandez Factory (1950) to the Bacardi Rum Factory (1960), an astonishing eighty-three works (thirty-nine of which were completed in 1955 alone), all variations on the thin concrete shell form of the hyperbolic paraboloid (hypar) for which he is so admired. The range of forms—from the subtle Cosmic Rays Pavilion at the National University of Mexico in Mexico City (1951; fig. 1) to the whimsical Lake Tequesquitengo signpost in Morelos (1957; fig. 2)—all have a rugged presence and exuberant freedom.

I met Felix Candela in 1983, when he taught my second class, at the suggestion of Mario Salvadori, after I had delivered a disastrous first lecture to my students at the Parsons School of Design in New York.

❸
Felix Candela. Sports
Palace. Palacio de los
Deportes for the
XIXth Olympics,
Mexico City

❹
Felix Candela.
Church of La Virgen
Milagrosa, Colonia
Vertiz Navarte,
Mexico City

At that time Candela and his wife, Dorothy, were living in the city, having emigrated from Mexico in 1971. He spoke to the class about his design and construction work, about the importance of direct observation and judgment, and about the limits of complex calculations and the natural resistance of shell forms. He described with relish how his design of the Palacio de los Deportes for the XIXth Olympics in Mexico (1968; fig. 3) was created and calculated in only a few weeks. The commission was assigned to Candela, Enrique Castañeda, and Antonio Peiri only eight months before the opening ceremonies. The initial approximate calculations by Candela were confirmed by detailed computations just days before the structure was successfully completed.

Candela delighted in the freedom and lightness that were possible with the geometric and structural order he had adopted. The syntax of the hypar forms allowed poetic expressions from the Antonio Gaudí-like chiaroscuro of the Church of La Virgen Milagrosa in Narvarte (1954–55; fig. 4) to the surreal Sales Office in Guadalajara (1960; fig. 5). Even when he took up steel, wood, and copper for the prickly Palacio de los Deportes, he applied the same simplicity and invention (figs. 6 and 7). The tension-stayed X-framed arches prefigure by a quarter-century the engineer Peter Rice's (1935–1992) own beautiful mastlike structures (e.g., the TGV Station roof in Lille, France, 1992, with the SNCF design bureau and the MOMI tent in London, 1992, designed with Future Systems of London). And, of course, the hypar form, even in this steel design, holds pride of place (fig. 7).

In early 1997 I contacted Candela with the idea of organizing a lecture series named in his honor to be held yearly at The Museum of Modern Art and at the Princeton University and MIT schools of

5
Felix Candela. Sales Office, Guadalajara. Candela is standing at the left.

6 and **7**
Felix Candela. Palacio de los Deportes under construction, Mexico City

architecture. The Museum lecture would be co-sponsored by the Structural Engineers Association of New York, an organization I founded with a group of New York structural engineers two years before. The idea of the lecture series had many sources. One was David P. Billington's theory of structural engineering as an art "parallel to but independent of architecture in the same way that photography...is parallel to but independent of painting."[4] As Billington argues in *The Art of Structural Design: A Swiss Legacy*, the key to understanding this art is to follow "the way in which mechanics and aesthetics are linked through practice."[5] Often, he claims, the art of structural engineering is misinterpreted along the lines of two competing theories.

> The first expresses the belief that all structure is merely a part of architecture and that the engineer's work is purely technical while the architect determines the aesthetics. The second holds that engineering is applied science and hence any aesthetic arises from the laws of nature, which dictate an optimum for each case.... Both ideologies are particularly damaging to the wider understanding of structural art in bridges because the first often implies that beauty requires great cost and thus little economic discipline, while the second suggests (erroneously) that beauty naturally occurs because efficiency automatically leads to elegance.[6]

Besides the Swiss structural engineers that Billington highlights in his book—Robert Maillart (1872–1940), Othmar H. Ammann (1879–1965), Heinz Isler (born 1926), and Christian Menn (born 1927)—the other structural engineers he acknowledges as artists of this kind include the Spaniard Eduardo Torroja (1899–1961), the Italian Pier Luigi Nervi (1891–1979), and, of course, Candela. Their arts are diverse, from the plasticity of Torroja's Coal Silo (1951; fig. 8) and Nervi's St. Mary's Cathedral in San Francisco (1970; fig. 9), both late works, to the minimalism of Maillart and Ammann. In the case of the work of Ammann,

8
Eduardo Torroja. Coal Silo, Instituto de Ciencias de la Construccion, Madrid

9
Pier Luigi Nervi. St. Mary's Cathedral, San Francisco

10
Robert Maillart. Schwandbach Bridge, Hinterfultigen, Switzerland

there is not a single line of aesthetic development—the George Washington Bridge (1931), the Bronx Whitestone Bridge (1935), and the Verrazano Narrows Bridge (1964), all in New York City, do not readily appear to be the work of the same artist. In fact the visual and intellectual strength of Ammann's work comes from the distinct individuality of each project. In other's work, like Nervi's and Candela's, the focus on and development of a particular formal aesthetic is more obvious. In Maillart's case one can follow the interplay of two distinct lines of development: the three-hinged arch (e.g., the Salginatobel Bridge; 1930; see page 129) and the deck-stiffened arch (e.g., the Schwandbach Bridge; 1933; fig. 10).

Besides the aesthetic qualities highlighted by Billington, what links the work of these engineers, from the particularism of Ammann to the dual research of Maillart and the narrow focus of Candela is a strong empirical focus, a direct concentration on things.

> —Say it, no ideas but in things—
> nothing but the blank faces of the houses
> and cylindrical trees
> bent, forked by preconception and accident—
> split, furrowed, creased, mottled, strained—
> secret—into the body of the light![7]

These well-known lines from the poem *Paterson* by William Carlos Williams capture the balance of aesthetic pleasure and objective attention that is present in the best of the art of structural engineering. Form is considered a category (indeed a constellation) of things, teased out by the arts of empiricism, practice, and observation.

While this applies well to the seven engineers presented in this collection of essays, there is also apparent in their work and words a struggle with "questions [that] are not answered by either empirical observation or formal deduction,"[8] questions that are philosophical and are subject to moral, social, or political inquiry, and reason. If "rationality rests on the belief that one can think and act for reasons that one can understand, and not merely as the product of occult causal factors which breed 'ideologies' and cannot, in any case, be altered by their victims,"[9] then the consciousness that emerges from engineering practice expands beyond the aesthetic and formal to these other questions. Through the connection to things, reason is then able to ask questions of fact and reach others through the physical expression of this reasoned inquiry.

The direct engagement with things that is characteristic of the work of Candela and these artist-engineers expresses human values broadly. This is clear from Candela's stark chapel in Lomas de Cuernavaca (1958–59; see page 168) and his Sales Office (fig. 5) in Guadalajara. The buildings address directly the question of what it is to be a man, a woman, in a particular time, going to church, going to work, or going to shop. The hyperbolic paraboloid form, the rough concrete work, the obvious evidence of hand crafting, the soaring, thin line of the shell's edge, all these touch, as Isaiah Berlin lists them: "The basic categories (with their corresponding concepts) in terms of which we define men—such notions as society, freedom, sense of time and change, suffering, happiness, productivity, good and bad, right and wrong, choice, effort, truth, illusion (to value them wholly at random)."[10]

Ove Arup's quote at the beginning of this essay is a clear allusion to romantic heroism—to the perfect act "in which freedom and self-fulfillment lie in the recognition by men of themselves as involved in the purposive process of cosmic creation."[11] Arup's admiration for Candela as the master builder echoed the Bauhaus "idea of creating a new unity of the welding together of many 'arts' and movements: a unity having its basis in Man himself and significant only as a living organism."[12] Arup shared with Walter Gropius this ideal of a "total architecture" and of the perfect work as a perfectly integrated organism. This ideal of totality requires the absolute authority of the work's author, a captain who will guide the ship of design to complete integration.

In one sense this idea of romantic heroism could apply to the man, Candela, who in 1936 joined the Republican cause "with enthusiasm,"[13] and who later, given the opportunity to build shells in Mexico, felt "as though all the previous events of my life began to make sense and to have meaning. I began to feel 'in form,' like an athlete, but mentally as well. I felt the moment had arrived to do something."[14]

But Candela's work belies this interpretation. The repetition of the hypar form stands apart from the architecture. As the critic Colin Faber wrote, "Once a structure stands, I believe it is dead to Candela. The forms are dropped and the last support is slammed away. As its forces start to play, the shell comes alive. It stands. That is all Candela needs to know and already, for him, 'it belongs in history.'"[15]

This is not total architecture. It is closer to the detached "stance" of the confident ironist, the observer and handler of "things." As Paz writes in his luminous essay on Marcel Duchamp's *Large Glass*, *Marcel Duchamp; or The Castle of Purity*, "Irony is the antidote that counteracts any element that is 'too serious such as eroticism' or too sublime like the Idea. Irony is the Handler of Gravity, the question-mark of *et-qui-libre?*"[16] Candela is closer in practice to the dialectical and antiretinal practice of Duchamp ("the beauty of indifference") than to the metamorphic and fertile heroism of Pablo Picasso. Rather than representing Arup's "balanced perfection," the work of Candela presents the concrete outcomes of a quizzical attention to things that actually suspends mastery. Engineering is not, as Billington correctly states, a simple reading or re-creation of nature. This is why the work of Candela continues to delight as well as serve us so well as a provocation to fresh thoughts about structure and architecture.

Constellations
Felix Candela was to have been the first lecturer in the series named for him, but in late 1997 he fell ill and on December 7 he died in Durham, North Carolina. Following Candela's wish, David Billington gave the first Candela lecture in April 1998, speaking on Candela's work. In the talk, Billington linked the work of Candela to those of other concrete shell designers and structural artists, including Anton Tedesko (1903–1994), Pier Luigi Nervi, and Heinz Isler. The essay by Billington and Maria Garlock that is included here is a further development of this lecture and serves to situate Candela alongside these peers. In 1999 the second Candela lecture was presented by the German engineer, Jörg Schlaich. After Schlaich the lectures were presented by Christian Menn (Switzerland) – 2000; Mamoru Kawaguchi (Japan) – 2001; Heinz Isler (Switzerland) – 2002; Leslie E. Robertson (US) – 2003; Cecil Balmond (UK and Sri Lanka) – 2004; and Stanford Anderson (US) on Eladio Dieste (Uruguay) – 2005.

The Candela lecturers were selected to "honor the most distinguished structural engineers active in design in the world today" by a committee that included Terence Riley (MoMA), Herbert Einstein and (until 2004) Stanford Anderson (MIT), Ralph Lerner, Stan Allen, and David Billington (Princeton) and, after 2000, Mutsuro Sasaki (Japan), Ricky Burdett (UK), and Antoine Picon (France). The Structural Engineers Association of New York was represented in the selection process each year by its president and president-elect. The lectures were intended as both an honor and an occasion for the engineers to present their work and consider the meaning of that work and its connection to, or distinctness from, the practice of architecture. Each talk was given with accompanying slides and, for the most part, in the informal narrative style that is usual in architecture and engineering project-based lectures. While the Museum venue was an explicit challenge to the engineers who were speaking, and to those in the audience, to reflect on the place of the engineer's work in the context of modern art, the speakers seized this opportunity to varying degrees, highlighting the extent to which they saw their work as art, invention, or problem-solving technique.

For this compilation, transcripts of the talks were expanded and modified by the speakers. Some of the earliest lectures have been updated and others were recomposed. The structures of the final essays—from first-person narratives of project histories to more theoretical frameworks—are an indication of each of the engineer's position with regard to the meaning of his work. The essays have been arranged in the reverse chronological order from that in which they were given. This order gives emphasis to Eladio Dieste (1917–2000), who was Candela's near-contemporary but whose work was, until the 1990s, less well known outside his native Uruguay.

Dieste's work, as Stanford Anderson describes it here and in his book, *Eladio Dieste, Innovation in Structural Art*,[17] is infused with the engineer's strong social convictions. As he wrote, "We, the nations of the third world, should not make the mistake of confusing the ends. Development is not an end in itself. Development will be beneficial as long as it is in accord with the ends of mankind and it will be detrimental if it forgets these ends."[18] He added, "When faced with the seduction of power, wealth and efficiency, without content, we must react."[19] For Dieste, the purpose of design is to enclose everyday life with lightness and natural light, employing local materials to achieve this end. Like Candela, he followed in the practical and poetic tradition of the Catalans Rafael Guastavino (1842–1908) and Antonio Gaudí (1852–1926),[20] joining the means of construction and the resistance of forms and materials in designs that also embody human hopes and values.

This tradition is evident also in the work of the Swiss engineer Heinz Isler (born 1926). Like Candela, Isler has concentrated his practice on thin concrete shells, inventing forms even freer and more playful than Candela's. Isler derives his forms not from analytical geometry (as were Candela's hypars) but directly from physical and funicular models – flexible membranes that assume the least energy, or minimal surface, for a specific boundary and force patterns. In the mid-1950s Isler invented two new form-making techniques, the first by using pneumatic models and the second by experimenting with hanging cloth models sprayed with water and put out to freeze in wintertime. Later, in 1965, he added a third technique that made shapes "by the flow method, by which the advancing velocity of a liquid inside a tube is varied. At the wall, velocity is zero because of friction, whereas in the center there is maximum velocity. A slowly expanding form leaves a square tube and forms a dome shape. This natural function produces lovely shapes."[21]

While the frozen cloths conform to the funicular[22] shape given by gravity, the other methods, pneumatic and flow, are hydraulic. Although constrained by physical processes, these experiments are not determined by gravity. The forms of projects like the BP Service Station in Deitingen (1968; fig. 11) and the Sicli Company Building in Geneva (1969–70; fig. 12) are adjusted for the site geometry, for the "loveliness" of their shape, and even the subtle interplay of light and shade that follows the necessity to curl the shell edge to prevent buckling.

Jörg Schlaich (born 1934) has been described by the architect Frank Gehry (born 1929) as the world's best living structural engineer. Schlaich, in his turn, has said that he thinks that Christian Menn is the world's greatest living bridge designer. Whatever else these opinions indicate, they point to the difference between Schlaich's team-based practice and the more solitary career of Menn. Alan Holgate's 1997 book on Schlaich, entitled *The Art of Structural Engineering: The Work of Jörg Schlaich and His Team*,[23] refers not only to his office and to his lab at the University of Stuttgart's Institute for Lightweight Structures (founded in 1964 by Frei Otto) but also to his collaborative relationships with architects. Schlaich worked first in the office of the great German engineering firm of Leonhardt und Andrä, where he helped design several telecommunication towers, and collaborated on the remarkable 1968 Munich Olympic Stadium roof. This membrane roof was designed by the architect Gunter Behnisch (born 1922) and a team that at first included Heinz Isler and then Frei Otto (born 1925), the great tensile-structures experimentalist, together with Leonhardt, Andrä, Schlaich, and Rudolf Bergermann. In 1980 Schlaich formed an independent practice with Bergermann.

The early work of Schlaich and his partners is classical in the sense that it is based on the evolution of existing types. Most obvious are

⓫
Heinz Isler. BP Service Station, Deitingen

⓬
Heinz Isler. Sicli Company Building, Geneva

13
Peter Rice, Renzo Piano, and Richard Rogers. "Gerberettes" for the Centre Georges Pompidou, Paris, in the Krupp casting factory in Germany

projects like the GRC Shell for the Stuttgart Federal Garden (1977), an explicit adaptation of Candela's Los Manantiales restaurant in Xochimilco (1958) by other material and construction means. Their cable net and glass grid structures are both developments of the Munich Olympic Stadium roof, using clamp connectors to join continuous, twinned tension cables or compression bars. The use of steel castings, facilitated by a close collaboration with German steelmakers, is another consistent element, applied by them recently to the twin lattice tower proposal of the THINK team for the redevelopment of the World Trade Center site in New York in 2003. Schlaich, like his contemporary and peer the Irish engineer Peter Rice (e.g., the Centre Georges Pompidou in Paris; 1972–76), uses cast steel joints in a conscious recovery of a late-nineteenth-century, particularly Victorian, technology, both for their expressive as well as functional potential. Rice, in particular, valued the *"trace de la main,"* or craftsmanship that is indicated by the roughness of the casting surface (fig. 13).

Like Isler's shells, the glass grid roof structures of Schlaich Bergermann und Partners are shaped to follow the funicular form delineated by the geometry of the supporting boundaries and the glass self-weight. In recent projects with Frank Gehry, such as the Berlin DG Bank Building and the proposed Museum of Tolerance in Jerusalem, the funicular form is further warped by Gehry's sculptural will. It is testimony to the openmindedness and curiosity of Schlaich and his partner in these projects, Hans Schober, that they are happy to adapt their glass shells to these improvisations with Gehry and his partners, Craig Webb and Edwin Chan.

The early bridges of Christian Menn are also evolutionary, drawing on the work of Robert Maillart for a series of elegant concrete deck-stiffened arch bridges beginning in 1957. In 1969 Menn designed his first prestressed concrete hollow-box beam bridge, the Salvanei Bridge. In 1971 he was appointed professor at the ETH (Federal Institute of Technology), where he had received his doctorate in 1956 with Pierre Lardy (1903–1958).[24] Since the 1970s, Swiss law has limited Menn to consulting on a few select projects, all of which have been concentrated and original works of bridge design.

In their later bridge work, both Schlaich and Menn have been open to original forms, with little regard for "taste." Marcel Duchamp once proposed "the search for 'prime words' ('divisible only by themselves and by unity')"[25]—an apt description of both men's bridge designs. Schlaich's Ting Kau Bridge in Hong Kong violates any precedents of "good" cable-stayed bridge design and for that reason remains indelible in one's memory.

Menn's great Felsenau (1974), Ganter (1980), and Sunniberg (1998) bridges are all of brilliant intelligence and raw simplicity. The towers of the Sunniberg Bridge in particular recall his fellow Swiss Alberto Giacometti's *Spoon Woman* (1926–27; fig.14)—enigmatic and curled. Even now, after years of familiarity, the Ganter Bridge remains iconoclastic.

Mamoru Kawaguchi (born 1932) is, like Schlaich, a great collaborator. Kawaguchi contributed to several of the key projects at Expo '70 in Osaka, including the space frame by Kenzo Tange (1913–2005) and the inflated Fuji Group Pavilion. Like Schlaich, Kawaguchi is also rooted in a strong lineage of engineer designers. He worked with the remarkable Yoshikatsu Tsuboi (1907–1990) and Tange on the cable-stayed roof of the Yoyogi Indoor Stadium for the 1964 Tokyo Olympics and with Arata Isozaki (born 1931) on the unfolding Palau Sant Jordi Sports Palace for the 1992 Barcelona Olympics. The 1998 Centennial Hall in Nara, also with Isozaki, even displays the deployment hinges as ornaments—an echo of the projecting construction corbels of the roman Pont du Gard in Nîmes, France. Kawaguchi, of all the engineers in this collection, is the most whimsical and humorous. He designs air arches, springing pantadomes, and flying fish. His exuberance adds a theatricality to his designs. Not only are many of the projects—for Expos, Olympics, and festivals—ephemeral, but their construction is an event, a happening, like the extraordinary Osaka Expo "Pavilion" of Billy Kluver (1927–2004) and others.[26] As did Candela, Kawaguchi delights in the dramatic moment "when the forms are dropped and the last support is slammed away.... It stands."[27]

Leslie Robertson (born 1928) is the only US engineer in this collection. He belongs to the two generations of US structural engineers that includes Fred Severud (1899–1990), August Kommendant (1906–1992), Paul Weidlinger (1914–1999), Fazlur Kahn (1929–1982), and William LeMessurier (1927–2007), whose originality found expression in collaborations with architects. Severud was associated with Eero Saarinen and Philip Johnson, Kommendant with Louis Kahn and Moshe Safdie, Weidlinger with Gordon Bunshaft and Marcel Breuer, Kahn with Myron Goldsmith and Bruce Graham, and LeMessurier with Hugh Stubbins and Helmut Jahn. Robertson has worked with Minoru Yamasaki (1912–1986), Wallace Harrison (1895–1981), Gunnar Birkerts

14
Alberto Giacometti. *Spoon Woman*. 1926–27. Bronze (cast 1986), 57 x 20¼ x 8¼" (144.8 x 51.4 x 21 cm). The Museum of Modern Art, New York. Acquired through the Mrs. Rita Silver Fund in honor of her husband Leo Silver and in memory of her son Stanley R. Silver, and the Mr. and Mrs. Walter Hochschild Fund

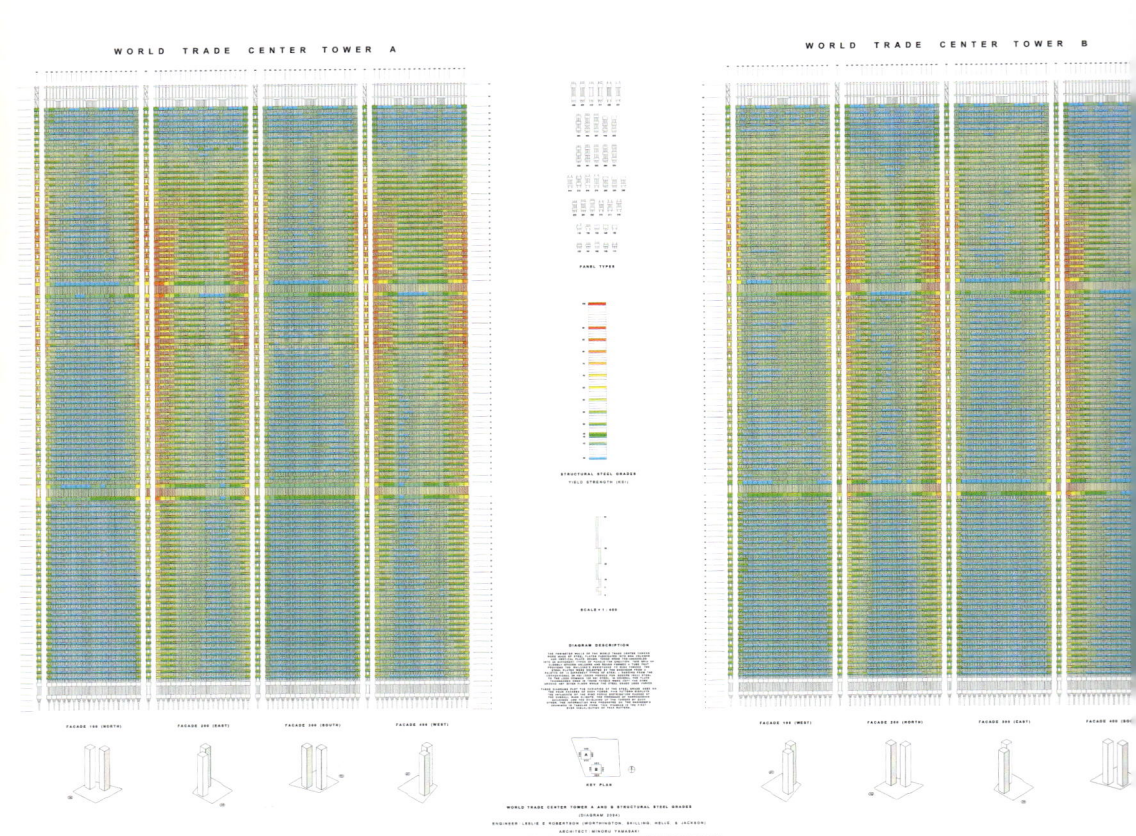

(born 1925) and, most notably, with I. M. Pei (born 1917). Robertson paints ideas with steel. On the World Trade Center towers (1970–2001) he used fourteen different steels (fig. 15) for the perimeter "tube" structure. By applying high-strength steel in some parts, he could use smaller plate thickness and thereby reduce their stiffness, to shed stress to the thicker, lower strength steel elements. The spectrum of steels used is the invisible expression of an integration, across all the possible extreme winds, measured by wind tunnel models of the surrounding urban terrain and the climatic idiosyncrasies (the "wind rose")[28] of the New York City area. This WTC design marked the culmination of similar uses of multiple steels on the IBM and US Steel Buildings in Pittsburgh.

Robertson's collaboration with Pei developed in the 1980s with the Bank of China Tower in Hong Kong (1990) and culminated in the remarkable bridge and tunnel at Miho, Japan (1996). The Bank of China Tower is particularly intriguing for the fact that the crystalline geometry of the outer surface disguises the offset geometry of the unique three-dimensional space frame structure of steel and concrete.

15
Guy Nordenson et al. World Trade Center Towers A and B structural steel grates (diagram 2004). 55" x 7'5" (139.7 x 226.1 cm). Guy Nordenson and Associates. This color-coded diagram shows the 14 types of steel used in the construction of the perimeter tube structure of the World Trade Center, as detailed in tables in the working drawings of Skilling, Helle, Christiansen, and Robertson, engineers.

This apparent flattening[29] of the structure to the minimal surface is reminiscent of both Mies and Severud's Seagram Building in New York (1958) and Bunshaft and Weidlinger's Beinecke Rare Book Library at Yale (1963), both instances where the structure is thinned to a surface of shadows and light (figs. 16 and 17).

Cecil Balmond (born 1943) is best known for his collaborative work with Rem Koolhaas (born 1944), Daniel Libeskind (born 1946), and Alvaro Siza (born 1933). Balmond is especially preoccupied with number and pattern. His first book, *Number 9*,[30] begins with the story of a boy mathematician faced with a riddle: "What is the fixed point of the wind?"

16 *(below)* Ludwig Mies van der Rohe, Philip Johnson, and Fred Severud. Seagram Building, New York. Section drawing

17 *(below, right)* Gordon Bunshaft and Paul Weidlinger. Beinecke Rare Book Library, New Haven, Conn.

To solve the riddle Enjil went to a secluded spot and sat in the shade of a banyon tree and blanked everything he knew out of his mind. The great blackness descended. Nothing moved. Shadows went into deeper shadows, layer into layer. A black disk grew. First as a dot, then a circle, then a rushing blind movement. Then the numbers came out, tumbling one over the other, rolling the patterns over in his head. There were star patterns, zigzags, squares, cubes, seesaws, and the weaving patterns going in and out, all twisting over each other . . . And there in the simplest patterns were points that did not move or change, no matter what the numbers were. And they were the fixed points.[31]

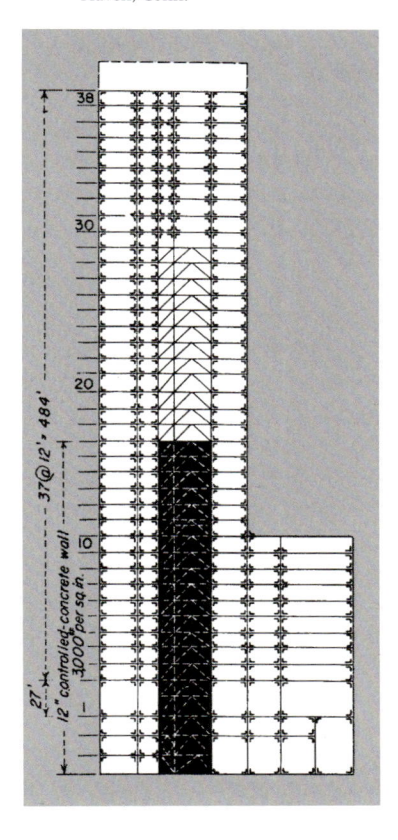

Though Balmond looks to manifest the patterns he discovers in his designs, he shares with Robertson the interest in the abstraction of structure. The spiral extension to the Victoria and Albert Museum (2003), designed with Libeskind, is an astonishing act of creative editing by Balmond that, like Ezra Pound's reduction of T. S. Eliot's draft of *The Waste Land*[32] to its final form, promises to be Libeskind's best work. The Bordeaux house (1998) and the Beijing CCTV Headquarters with OMA and Rem Koolhaas (under construction) both elaborate and display the structure as ornament—an ambiguous and ironic use of the structure's usual implications of authenticity, while the delicate, stressed, concrete sheet of the Portuguese Pavilion for the 1998 Lisbon Expo, with Alvaro Siza, and the later tensile roof of Braga Stadium (2000–2004), with Eduardo Souto de Moura (born 1952), link Balmond's example to the historical lineage of Candela's funicular abstraction.

Economy

> UNE CONSTELLATION
>
> froide d'oubli et de désuétude
>
> pas tant
>
> qu'elle n'énumère
>
> sur quelque surface vacante et supérieure
>
> le heurt successif
>
> sidéralement
>
> d'un compte total en formation[33]

The word "economy" has a vast etymology that extends back through the Greek words for "home" (*oikos*) and "law" (*nomos*) to the more contradictory terms of parish and village, number and nomad. Economy is central to engineering and its aesthetic. Candela's shells are reductive, not as total works of art but as simplification. For the Greek astronomers and the writers and scientists who followed them, the ideal was the cold purity of the patterns apparent in the stars and constellations and their motions. The poem of Stéphane Mallarmé (1842–1898), *"Un Coup de dés jamais n'abolira le hasard,"* has been a modern icon of this ideal ever since Mallarmé showed the original manuscript to his friend, Paul Valéry (1871–1945), because it is as cold, impenetrable, and beautiful as the stars. Like all great poems its economy is also absolute. Admirers of *"Un coup de dés,"* from Valery to T. S. Eliot and Marcel Duchamp, have been drawn, I think, to the absoluteness of sublimation that the work presents. There is a kind of mysticism to it as well. It is to use other oppositions—Apollonian, classical, not Dionysian or romantic.

Type is at its root a kind of constellation, and engineering is necessarily typological. That is not to say there is no iconoclastic structural art. In fact, we see in the work of some of the engineers in this collection clear examples of this. But the iconoclasms are not instances of "cosmic creation" as Berlin describes those of the great romantics. They are instances of economic discovery, the uncovering of some new laws of making things that spring from a particular culture or situation—hypar shells in Mexico, frozen shells in Switzerland, masonry vaults in Uruguay, steel castings in Germany, multiple varieties of U.S. steel, and highly industrialized structural machines in China and Japan. The types that emerge are works of art in the realm of things and orders, not feeling and self-expression.

This idea of type as constellation can also be explained by reference to the difference between the Brooklyn Bridge (1888), and the *Statue of Liberty* (1884). The Brooklyn Bridge is much more than just a bridge, as its utilitarian neighbors, the Manhattan and Williamsburg bridges, make clear. The ephemeral deck truss hung from the parabolic main cable and radiating stay cables contrasts with the thick granite and gothic towers to embody both heaven and earth—the same opposites that charge Mallarmé's poem. The bridge is perfectly proportioned and unforgettable. The structure designed by Gustave Eiffel (1832–1923) for the *Statue* is, on the other hand, revealed only on the inside, and piecemeal at that (fig. 18). It is effectively invisible, and yet once it is understood through drawings and photographs (fig. 19), it is of astonishing originality and *delicatesse*. It is an idea caught lightly, invisibly, in a thing.

"Thus structure, the intangible concept, is realized though construction and given visual expression through tectonics,"[34] Eduard F.

18 Gustave Eiffel and Auguste Bartholdi. Interior of the *Statue of Liberty*, showing the secondary iron frame and armature supporting the copper skin

19
Gustave Eiffel.
Preassembly of the
structure for the
Statue of Liberty
in Paris

Sekler writes, "[and the] atectonic is used... to describe a manner in which the expressive interaction of load and support in architecture is visually neglected or obscured."[35] Or referring to a verse by Henry Wadsworth Longfellow that Ludwig Wittgenstein claimed as his motto:

> In the elder days of art
> Builders wrought with greatest care
> Each minute and unseen part,
> For the Gods are everywhere.[36]

This economy is a manner of conscience and practice, that is, a series of choices made on the belief "in the intelligibility of the notion of objective inquiry ... [and] the discipline required by the dedication to the ideal of *correctness*."[37] Tectonics is an expression of the economy of design that pays homage to the economy of evolution—and survival— "with greatest care."[38] The atectonic rather sublimates the structure into ornament, or just nothing. Construction itself can offer, on occasion, a time for theater and performance that may echo in the cultural memory of the built thing.

The works of the engineers that are represented in this collection divide into two general categories. Dieste, Isler, and Menn are structural artists; each had or has an autonomous practice that transforms the meaning of the art. They invent and develop forms that have a clear structure, are constructed in a direct and legible way, and "give visual expression" to these facts through their form and details. Like the Brooklyn Bridge, they are made of what you see, even if they are open to multiple readings. Their "engineering . . . is a fairly anonymous art"[39] since they are in public use and do not carry a signature. But each has a distinct way of working and has developed a new formal language that has changed the history of the art for those that practice it.

Balmond, Kawaguchi, Robertson, and Schlaich are not visual artists in the same way as the other three, but that only makes their work more challenging to think about. This is where the example of Duchamp is instructive. "Structure, the intangible concept" and the "unseen part" that are "realized through construction" could just as easily be a description of the "Ordre des 15 operations de montage general" for his *Etant donnés: 1e la chute d'eau, 2e le gaz d'éclairage*[40] (fig. 20), as indeed it could be for that other prefabricated lady—the *Statue of Liberty* (fig. 19). These four engineers always work in collaboration with architects. While Balmond and Schlaich have developed geometric patterns and structural ideas that persist across different collaborations,[41] these four mostly apply a hidden hand. These degrees of anonymity or impersonality lend, for those sufficiently curious and willing to research, a suspense and even (in the case of Kawaguchi) a sense of theater and humor to their work. The invisible creativity of Robertson's World Trade Center structural design is, especially after the tower's destruction, as poignant and memorable as any great work of conceptual art.

20
Marcel Duchamp. *Étant donnés: 1e la chute d'eau, 2e le gaz d'eclairage*. 1946–66. Mixed-media assemblage, approx. 7' 11½" high, 70" wide. Philadelphia Museum of Art. Gift of the Cassandra Foundation, 1969

Notes

1. Ove N. Arup, "Foreword," in Colin Faber, *Candela: The Shell Builder* (New York: Reinhold, 1963), 8.
2. Octavio Paz, "The Heroic World," in *The Bow and the Lyre*, trans. Ruth L. C. Simms (New York: McGraw Hill, 1973), 183.
3. Faber, *Candela: The Shell Builder*, 12.
4. David P. Billington, *The Art of Structural Design: A Swiss Legacy* (New Haven: Yale University Press , 2003), 13.
5. Ibid., 15.
6. Ibid., 14–15.
7. William Carlos Williams, "The Delineaments of the Giants," in *Paterson: Book I*, rev. ed. (New York: New Direction Publishing, 1992), 6-7
8. Isaiah Berlin, "Does Political Theory Still Exist?" in *The Proper Study of Mankind* (New York: Farrar, Straus and Giroux, 1998), 84.
9. Ibid., 89.
10. Ibid., 83.
11. Ibid., 72.
12. Sigfried Gideon, *Space, Time and Architecture*, 5th ed. (Cambridge: Harvard University Press, 1967), 511.
13. Faber, *Candela: The Shell Builder*, 12.
14. Ibid., 13.
15. Ibid., 77.
16. Octavio Paz, *Marcel Duchamp; or The Castle of Purity*, trans. Donald Gardner (London: Cape Goliard, 1970), 31.
17. Stanford Anderson, ed., *Eladio Dieste, Innovation in Structural Art* (New York: Princeton Architectural Press, 2004).
18. Eladio Dieste, "Tecnica y Subdesarrollo," in Antonio Jimenez T et al. eds., *Eladio Dieste 1943–1996* (Montevideo: Dir. General de Arquitectura y Vivienda, 1996), 262.
19. Ibid., 263.
20. Edward Allen, "Guastavino, Dieste, and the Two Revolutions in Masonry Vaulting," in Anderson, *Eladio Dieste*, 66–75.
21. Billington, *Art of Structural Design*, 139.
22. Funicular, as in *funis* + *-iculus*, or little rope, is used to describe the shape of a suspended string or net of strings. Also *funiculi', funicula'*!
23. Alan Holgate, *The Art of Structural Engineering: The Work of Jörg Schlaich and His Team* (Stuttgart: Edition Axel Menges, 1997).
24. See Billington, *Art of Structural Design*, in particular 169–72, for a discussion of the key role of Pierre Lardy.
25. Marcel Duchamp, in Paz, *Marcel Duchamp; or the Castle of Purity*, 10.
26. Billy Kluver et al., *Pavilion / Experiments in Art and Technology* (New York: E.P. Dutton, 1972).
27. Faber, *Candela: The Shell Builder*, 77.
28. The "wind rose" is the diagram of the fastest winds measured for a particular time-exposure period over the cardinal coordinates. The "petals" are the directions of prevailing wind—in New York, the Northeast and Northwest mostly. See http://www.wcc.nrcs.usda.gov/climate/windrose.html.
29. "No terms taken from other art—whether from antecedent paintings or from Picasso's own subsequent Cubism—describe the drama of so much depth under stress. This is an interior space in compression, like the inside of pleated bellows, like the feel of an inhabited pocket, a contracting sheath heated by the massed human presence." Leo Steinberg, quoted by Rosalind Kraus, in "Flattening Space," *London Review of Books* 26, no. 7 (April 1, 2004).

30 Cecil Balmond, *Number 9: The Search for the Sigma Code* (New York: Prestel, 1998).
31 Ibid., 19.
32 T. S. Eliot, *The Waste Land: A Facsmile and Transcript of the Original Drafts*, ed. Valerie Eliot (New York: Harcourt Brace Jovanovich, 1971).
33 Stéphane Mallarmé, *Un Coup de dés jamais n'abolira le hasard* (Paris: Gallimard, 1914).
34 Eduard F. Sekler, "Structure, Construction, Tectonics," in Gyorgy Kepes, ed., *The Structure of Art and Science* (New York: George Braziller, 1965).
35 Eduard F. Sekler, "The Stoclet House by Joseph Hoffmann," in *Essays in the History of Architecture Presented to Rudolf Wittkower* (London: Phaidon Press, 1967), 230–31, quoted in Kenneth Frampton, *Studies in Tectonic Culture* (Cambridge: MIT Press, 1995).
36 Longfellow as quoted in Harry G. Frankfurt, *On Bullshit* (Princeton: Princeton University Press, 2005), 20. "Wittgenstein once said that the following bit of verse by Longfellow could serve him as a motto."
37 Frankfurt, *On Bullshit*, 65.
38 Cf. the sculptor Richard Serra, "Since I chose to build in steel it was a necessity to know who had dealt with the material in the most significant, the most inventive, the most economic way," in Hal Foster, "The Un/making of Sculpture (1998)," in Hal Foster, ed., *Richard Serra* (Cambridge: MIT Press, 2000), 186.
39 "Until lately art has been one thing and everything else something else. These structures are art and so is everything made. The distinctions have to be made within this assumption. The forms of art and of non-art have always been connected: their occurrences shouldn't be separated as they have been. More or less, the separation is due to collecting and connoisseurship, from which art history developed. It is better to consider art and non-art one thing and make the distinctions ones of degree. Engineering forms are more general and less particular than the forms of the best art. They aren't highly general though, as some well designed utensils are. Simple geometric forms with little detail are usually both aesthetic and general." Donald Judd, review of *Twentieth Century Engineering* exhibition at The Museum of Modern Art, 1964, in *Donald Judd: Complete Writings 1959–1975* (New York: New York University Press, 1975), 138.
40 Marcel Duchamp, *Manual of Instructions for Marcel Duchamp, Étant donnés: 1e la chute d'eau, 2e le gaz d'éclairage* (Philadelphia: Philadelphia Museum of Art, 1987).
41 The art-historical technique of using side-by-side slides as a means of interpreting artworks is useful when studying the works of engineers who act in collaborations, such as Peter Rice, August Kommendant, or Paul Weidlinger, as well as the four included here. Renzo Piano, with and without Rice, Louis Kahn, with and without Kommendant, or Gordon Bunschaft, with and without Weidlinger, all do very different work, as do these same engineers with other collaborators (e.g., Richard Rogers, Moshe Safdie, and Marcel Breuer, respectively).

The Felix Candela Lectures

❶
Massaro Agroindustries,
Joanicó, Uruguay. Loading area

Eladio Dieste: A Principled Builder

STANFORD ANDERSON

Eladio Dieste took the construction method of reinforced masonry into radically new territory. A single image suggests Dieste's accomplishment and the reasons that his work deserves attention. At the fruit-packing plant of Massaro (1976–80; fig. 1), a daring set of double-cantilever vaults protects the area where trucks deliver and collect large quantities of agricultural produce and materials—just a way station in a humble process. Most of Dieste's work was for agricultural or industrial purposes, and yet almost every building has true architectural merit. The technology is exceptional; while reinforced masonry had existed in simple ways for a long time, the use of reinforced masonry in an inventive manner is basically the contribution of Dieste. Some idea of his achievement can be gleaned from this image: the three vaults span 12.7 meters laterally and are cantilevered 13 meters in both directions from the single line of four columns. The vaults are very thin, only one layer of brick (10 centimeters thick). We immediately confront an extraordinary technical feat of building: a large covered area with a very thin vault of long, double cantilevers.

Dieste was born in 1917 in Artigas, Uruguay. Schooled at the university in the capital city of Montevideo as an engineer, he never claimed to be anything other than an engineer. Yet most of his works deserve to be understood as architecture. This portrait of Dieste (fig. 2), seated at home, was not taken on the day I met him, but it gives the same impression. He was already suffering a degenerative disease, but maintained his spirit and continued to work until his death in 2000. On the first day that I saw a few of Dieste's buildings, I was taken to see him in the evening. Genuinely enthusiastic from the experience of the day, I extolled the qualities of his buildings.

2
Eladio Dieste, in his home, Montevideo, Uruguay, 1996

Dieste looked at me with the same twinkle as can be seen in the photograph and admonished me: "I too follow the laws of physics."

Of course, what alternative is there but to follow the laws of physics? But important points are concealed in this obvious statement. With Dieste there is no willful invention of form; there are rather innovative forms that follow from first principles. Dieste would never have realized his unprecedented works if he had worked according to the conventions of society or of engineering production. The technical characteristics of these buildings were not covered by any code or standard. He had the courage to do what had not been done before. The reason he had courage was that he had been very well schooled in principles starting from his fundamental knowledge of science.

There are several reasons to attend to Dieste's work. He was a principled builder, who deserves a place in the realm of architectural knowledge, beginning with his innovations in structure. Dieste took up the very ordinary material of brick—perhaps the earliest building material other than twigs and stones—and raised it to a completely new level. Through this achievement with something as ancient and well known as brick, his work suggests the possibility for innovations of equal magnitude hidden in other traditional materials, or, for that matter, in new materials. Dieste thus offers a model of innovation.

Beyond the simple material, Dieste's work employs simple construction techniques and, in many of the buildings (there are some exceptions), the simplest of craftsmanship. It is plausible then to think that Dieste's work could be of interest in other societies that have even less in the way of resources—fewer material resources and limited construction techniques. There may well be places in the world where the direct application of Dieste's ideas is still a fruitful possibility. Learning from Dieste need not stop there. Today, in the schools and practices of technologically advanced societies, methods of representation overwhelm production. Dieste's example does not direct us to neglect the new, but rather to adopt it under rigorous inquiry.

Finally, I return to the principled builder again. This is a man of high ethical standards, deeply concerned about society. He understood engineering and architecture as profoundly implicated in one another and embedded in culture and in ethics. There is a wholeness to his personality and his work that is a model whether one aspires to be a painter or a writer, or anyone who seeks to be productive and ethical in order to contribute to the well-being of society.

Turning to his work, Dieste developed two types of reinforced masonry vaults, one of which he called "self-carrying vaults," as at Massaro Agroindustries in the countryside north of Montevideo (fig. 3). The long vaults for the storage and work area open to the vaults of the loading and office area in the distance. The vaults are prestressed to take bending forces and thus act as beams. Unlike conventional barrel vaults, such vaults do not require continuous support in the walls or in the tympanum. Here 10,000 square meters are covered using only

❸ Massaro. Main warehouse to loading area

❹ Massaro. Detail of vaults

twenty-four columns. The small number of point supports facilitates both circulation and an array of alternative sources of natural light. Additionally, Dieste could meet functional needs that would be precluded in buildings requiring continuous supports.

In the long storage area of Massaro, there are two groups of self-carrying vaults, end to end, each five vaults wide. The longitudinal span between supports is 35 meters. In the three middle bays, the two groups do not meet, and thus provide light in the center of the space. The vaults are also perforated for light. The exterior protective wall of this building is not a bearing wall. Perimeter light below the vaults would have been possible, as seen in other buildings of this structural type. Inside, there is a long span between supports and the 16.5-meter cantilevers (see the middle of figure 3). The narrow valley at the meeting of the vaults is remarkable as the vaults span between distant columns.

The drama of the ends of the main vaults cantilevering over the double-cantilever entry vaults is apparent in figure 1; the radical thinness of the vaults can be seen in figure 4. Such vaults have steel reinforcement between the bricks, loops of prestressing steel and steel mesh in the crowns of the vaults, and a cement parging that secures the bricks and steel and also provides the only roofing material. The cross sections of Dieste's vaults are always catenary curves, the most efficient structural form. When these vaults meet laterally, their opposing thrusts cancel one another and result in the economical, narrow valleys between vaults. When the vaults terminate, the lateral thrust must be resisted. Here, too, Dieste found an innovative and economical solution (fig. 5). Rather than a wall or any kind of continuous support, Dieste uses a lateral edge beam. The edge beam is small at the ends, where it receives the load from a small part of the vault, but widens toward the center as it sustains increasing force. The vertical and horizontal thrusts are finally resolved in a column, triangulated against the lateral force. The plans and sections of Massaro (figs. 6, 7) reveal the extraordinary structural economy of this large construction.

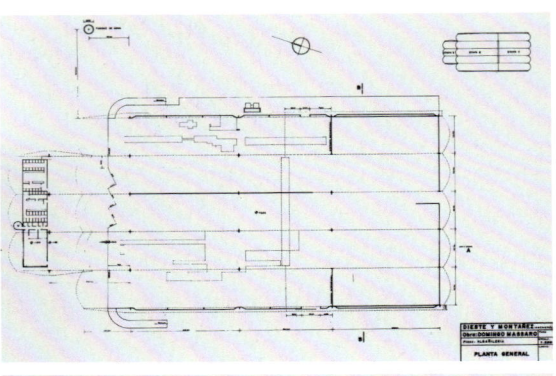

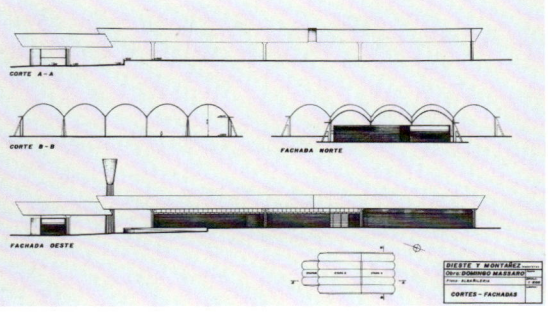

The municipal bus terminal in Salto, a provincial city in the north of Uruguay, also employs double-cantilever vaults on a single line of columns (1971–74; fig. 8). The cantilever is 12.13 meters on either side, an instance where the cantilever is used for the specific functional purpose of giving full protection to a bus. Dieste was always concerned with how form gives strength. A simple plane provides little strength. Give it a curved surface, and strength is developed. Enhancing this simple principle by using structurally sound catenary cross sections and prestressed reinforcement allowed Dieste to create these highly efficient vaults acting as beams.

Structural details of the Salto bus terminal appear in figure 9. The partial plan at the left shows the reinforcement through the crown of the vault; this prestressing was accomplished in the simplest possible manner. The drawing shows the prestressing steel in its final configuration. The process was as follows: after the masonry of the vault was in place and still supported with formwork, the workmen laid down loops of steel—a long, narrow loop at the center of the crown of the vault, a longer loop around that, and another loop around that, as many loops as Dieste's calculations required; the ends of the loops were then anchored to the bricks, as indicated by the horizontal lines in the drawing; using a simple carjack, the loops were then pinched together at the middle, stretching the steel and developing the prestressing.

5 *(left)*
Massaro. Detail of vaults and edge beam

6 *(above)*
Massaro. Plan

7 *(below)*
Massaro. Sections and elevations

34 | THE FELIX CANDELA LECTURES

Sophisticated equipment not being available to him, Dieste found a means that would quickly and easily provide the required prestressing.

Rather than cantilevers, the special feature of the maintenance hangers for the metro system of Rio de Janeiro is the provision of generous light throughout a very large work area (52,000 square meters; 1971–79; fig. 10). Here Dieste perforated the vaults, but, more significantly, he alternated the height of adjoining self-carrying vaults, thus creating large clerestory windows throughout the workplace.

For a tour de force, consider this structure (in brick!) that cantilevers four directions off a single column (1975–76; fig. 11). The "Sea Gull" was built as protection for the pumps at a service station in Salto. In Dieste's work, this is a rare divergence from true structural principles. This is no longer a vault, but a beam that requires heavy reinforcement, especially in the central valley. But there is a logic here, too, providing maximum access with minimum support.

(below) ❽
Municipal bus terminal, Salto, Uruguay

(right) ❾
Municipal bus terminal, Salto. Sections, reinforcing plan, and details

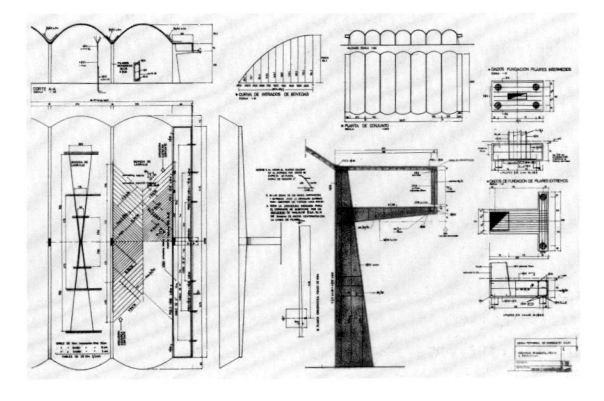

❿ *(left)*
Rio metro maintenance hangar, Rio de Janeiro, Brazil

⓫ *(above)*
Barbieri e Leggire Service Station, Salto. 1975–76; relocated at the south approach to Salto, 1996

12
TEM factory, Montevideo, Uruguay. Interior

The second vault type that Dieste used repetitively is the "Gaussian" vault, a vault of double curvature. The TEM factory in Montevideo (1958–62; fig. 12) is an early example of a discontinuous double-curvature vault: double-curvature because of the changing S-section along the catenary arcs of the vault, and discontinous because each vault is complete in itself, except where it joins a neighbor at the springing. The discontinuity provides for skylights throughout the covered area. Double-curvature vaults are typically used when a very long span is required in the lateral dimension—in this example, 43 meters. In contrast, self-carrying vaults can provide a very long span in the longitudinal direction; theoretically, they could also provide a long span laterally, but then the height would be very great. A large lateral span with a relatively low rise presents the problem that the vault is prone to buckle, break, and collapse. If the problem were to be solved by more and heavier material, an inherent negative feedback would result. That way the problem is solved only by brute force, an approach that Dieste would never take. Instead, he developed strength in the section through shape—the double curvature. We are returned to Dieste's principle that form is the critical generator of efficient structure, but also of good architecture.

The Port Warehouse in Montevideo, the reconstruction of an earlier warehouse damaged by fire, has a span of 50 meters, the longest such span in Dieste's work (1976–79; fig. 13). Dieste respected the quality of the old building with its historic brick walls (fig. 14), and also saw the economy in retaining these walls. From the exterior, the existing walls and the curvature of the new vaults are apparent, but the interior reveals the quality of space and light gained with double-curvature vaults (fig. 15). Dieste, working under conditions that he also helped to create, found reinforced masonry was the most economical way to build. With this technique he achieved welcome visual characteristics and, most important, remarkable qualities of light. Within these spaces, if one is not looking toward the skylights, there is an even stronger sense of how light diffuses through the space (fig. 15). In the Port Warehouse, the old walls provide continuous support, but Dieste's

(above) ⓭
Port Warehouse,
Montevideo. Interior

(below) ⓮
Port Warehouse.
Exterior

(right) ⓯
Port Warehouse.
End wall and vaults

double-curvature vaults can be carried on rather widely spaced columns, as can be seen in the TEM factory interior (fig. 12).

The longitudinal section of the Port Warehouse clarifies the system of Dieste's double-curvature vaults (fig. 16). At the top of the old brick wall is a new reinforced-concrete beam that ties the wall together, and establishes a uniform height and span to facilitate construction and assure a regular structure. The vault springs from the level beam, such that the first bricks are in a straight line. A movable formwork allows the construction of each unit of the discontinuous vaults (fig. 17). The formwork is shaped to change from the flat springing to the pronounced S-shape at the crown of the vault. Between those two points, there is a continuous transition from the straight line to the S-curve. In the transverse direction, every line that can be drawn is a catenary curve of a different amplitude, yielding not only the S-curve, but also the low and high terminations of each unit of the vault, and thus the opportunity for light.

In all his work, Dieste invented efficient construction techniques. With concrete, the formwork has to remain in place for days as the concrete cures. With brick, there is little wet mortar, and the bricks are already stable. Consequently, Dieste could build one of these vaults in a day, leave it overnight, then drop the form and move on to the next

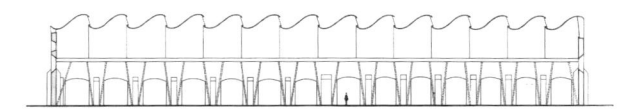

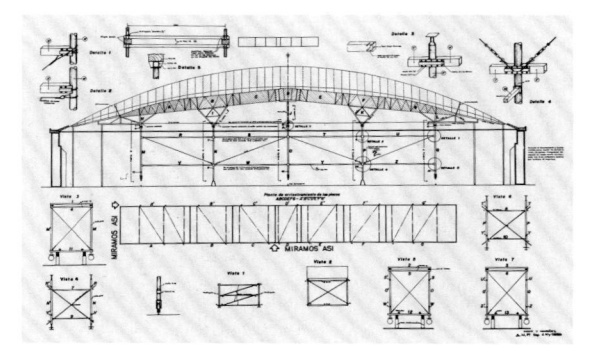

⓰ *(above)*
Port Warehouse. Longitudinal section
⓱ *(left)*
Port Warehouse. Formwork details
⓲ *(below)*
Port Warehouse. Vault section

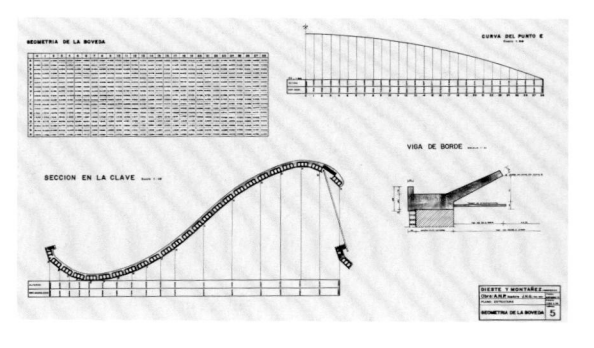

bay the following day. Because of this quick process, the mortar could still be tooled to give the good finish of the underside of the vaults. The construction process was always one of Dieste's concerns, for he was a builder. He looked for economy in structural concept, in materials, and in construction, both as a matter of principle and because he had to win commissions that involved economic competition. The detail section at the crown of one vault (fig. 18) shows that the construction is only one brick thick. Here the brick has some dimension as it is shaped and has cavities, but it is still only approximately 12 centimeters thick.

Dieste wrote, "If I had to synthesize what has driven our search, I would say that it is the perennial value of the surface itself." "Surface" for most of us is a rather dangerous word. It seems to suggest things like "surface treatment" and "surface coating" or superficiality. But clearly Dieste understood something truly fundamental in the idea of surface. He recognized that the surface offered a realm of formal exploration that could, in turn, solve structural problems. Again, solving the problem of buckling in the Gaussian vaults, giving strength in the center of these spans, is solved by the shape that is given to them. Dieste always looked for the way in which the forming of space, the shaping of the volumes, the use of the surface, would give economy and efficiency to his structures.

Dieste rejected reliance on rectilinear frame systems, but he also resisted structural solutions that relied on two-dimensional curved forms such as arches and ribs. Such approaches were an invitation to solutions based on additional material rather than on structural efficiency. To the contrary, Dieste's structural innovations relied on the efficacy of surfaces with particular formal properties. The simple catenary curve of a reinforced self-supporting vault allowed it to perform as a beam. The S-shaping of each band of Gaussian vaults gave it the stiffness to span great distances. Of course, these vaults are of material and have thickness, but their shape is still more fundamental to their capacity. Dieste wanted us to know this. Another look at figure 4 readily reveals that the physical facts of Dieste's buildings directly convey his idea of "the perennial value of the surface itself."

Dieste was, then, a master of structure and construction, but other qualities are also recurrent: the proportions of whole and part; the economy and elegance of the materials; the detail of the parts; and, above all, the knowing use of light as it plays on, and especially as it is admitted into, these buildings. These are the qualities of the work of a fine architect.

Brick. That Dieste built in brick is not a matter of nostalgia, but it is no doubt a significant aspect of why Dieste's work has been comparatively neglected in the historical and critical literature. Why should we allow historiographies that honor so-called modern materials to have a privileged place in our thinking when Dieste demonstrates that traditional materials can be used so innovatively? Dieste won the opportunity to build so much because he could build more economically than those who would use steel or reinforced concrete. Structural economy and rapidity of construction are not matters of nostalgia. Dieste made still other claims: about thermal qualities, aural qualities, and more. Consider that the color photographs presented here were taken two years ago of buildings that are thirty, forty, and almost fifty years old. It is evident that the quality of the material as it gives effect in terms of space, light, surface, and texture is very positive, whereas we know so many buildings of exposed concrete that, in a shorter time, have had great difficulties. Yes, Dieste was also aware of the charm, the visual quality, the human scale of bricks. Still, it is not nostalgia that his buildings evoke, but the sum of the very positive qualities of a material that could be supported on fundamental grounds.

Central to my concern with Eladio Dieste is to emphasize, more than others before me, that Dieste was not only an engineer but also an architect. Despite Nikolaus Pevsner's famous dismissal of bicycle sheds in evoking true architecture, I consider everything illustrated so far—Dieste's factories, warehouses, and other utilitarian structures—to be architecture. Dieste also had a few opportunities to work in building types that are unqualifiedly recognized as architecture. We now turn to three such: a house and two churches.

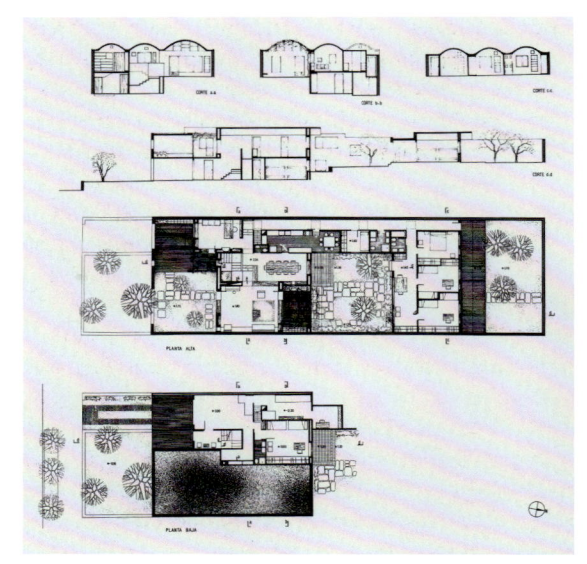

19 *(left)*
Dieste house, Montevideo. Plans and sections

20 *(below left)*
Dieste house. View from top of entry stair to sitting area and courtyard

21 *(below)*
Dieste house. View from sitting area to dining area and courtyard

Dieste built a house for his family in a pleasant part of Montevideo, overlooking the broad La Plata River that separates Uruguay and Argentina. The long, narrow site (12 x 50 meters) is fully occupied with the house and its consequential open spaces: the front area, raised front terrace, internal courtyard, and back garden (fig. 19). From the lower level entrance of the house, one rises to a point that is effectively the center of the house: the sitting area opening to terrace and courtyard (fig. 20), and the dining area a few steps above (fig. 21), with the kitchen and private areas beyond. The house is early in Dieste's work (1961–63) and early in the development of the self-carrying vault. The vaults are small (4.25-meter spans). As seen in the relation between the sitting and dining areas (fig. 21), the form of the vaults gives identity to each of the places, but Dieste's self-carrying technique also gives continuity between them. The vault of the sitting area continues over the terrace, perforated to provide a transition of space and light. The heavy, cellular quality of Le Corbusier's earlier and comparable Maisons Jaoul at Neuilly-sur-Seine, France (1954–56),which also employ tie-rods, provides a telling contrast to the lightness of Dieste.

Also early in his production, Dieste was afforded the opportunity to build a church in a small community of agrarian and manual laborers. The Church of Christ the Worker in Atlántida (1958–60; fig. 22) was intended as a simple construction, but Dieste made of it an extraordinary work. The interior of the church possesses remarkable qualities of light and space (fig. 23). Dieste's religious position is also on view in the close relationship of the congregation to the priest—and this before the mandates of Vatican II. Viewed from the exterior (fig. 22), the continuous double-curvature vault of the roof (maximum 18.8-meter span) begins over the main entrance, supported by the undulating walls composed of ruled surfaces. In the sections (fig. 24), it is apparent, but surprising, that these two curved surfaces, wall and roof, meet in a level plane. The sections also show that the low point of the vault is flat; in this way, Dieste was able to place a tie-rod within the vault itself and take up its forces in the edge beams at the top of the walls. An image from the late stages of construction (fig. 25) shows these structural features, and dramatizes the fact that the building is structurally sound without the end walls. From inside, looking toward the entrance wall (fig. 26), Dieste notes the structural independence of

(right) **22**
Church of Christ the Worker, Atlántida, Uruguay

(below) **23**
Church, Atlántida. Interior view toward altar

(right) **24**
Church, Atlántida. Sections

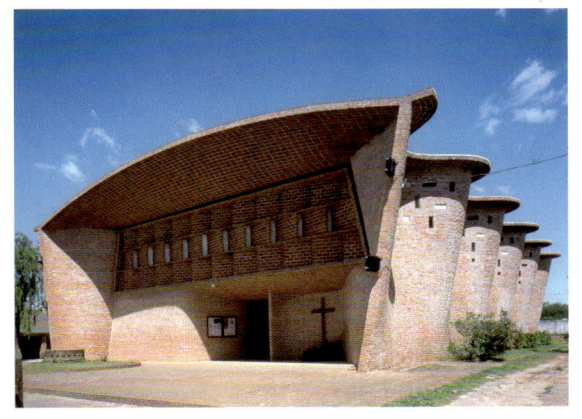

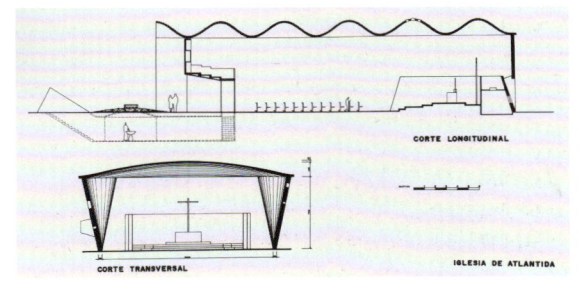

25
Church, Atlántida. View in late stages of construction

26
Church, Atlántida. Interior view toward entrance wall

this wall with a line of light around its edge and with the obviously nonstructural louvers for indirect light. The other main source of light is the penetrations in the reverse curves of the walls. Thus none of the sources of light are apparent to a person entering this church with its luminous surfaces and space (fig. 22).

An opportunity that became another remarkable church by Dieste was occasioned by the destruction of the early-nineteenth-century Church of San Pedro in the provincial town of Durazno (1967–71; figs. 27, 28). The narthex that supports the high bell tower of the facade survived, but the entire basilical nave and its roof were lost. Dieste transformed the commission for a simple reconstruction into what I claim to be one of the great works of architecture of the late twentieth century (fig. 29). Light diffused throughout the space makes the exquisite brickwork almost radiant, culminating in the burst of light over the altar. One is first absorbed with the experience of light, space, and craftsmanship. Then more practical questions present themselves. How is it that the side aisles, unlike other basilicas, are completely open rather than separated by a row of columns? How can there be a continuous light source between the wall and the roof? The answers to these questions are best revealed in an axonometric section (fig. 30).

Contrary to normal expectations, the structure does not span across the width of the church. Rather, the nave walls are actually very thin, but densely reinforced, high beams that span 32 meters from the narthex wall to the presbytery. The roof is a thin, folded plate construction in brick (7.8 centimeters thick), also spanning the length of the church. The small steel studs between the wall and roof are not bearing elements, but rather transfer wind forces between the wall and roof. It is the structural daring of the church that allows the unanticipated and effective lighting. Turning to leave, one is surprised by another source of light, a rose window at the inner wall of the old narthex—a rose window in brick, floating in space (fig. 31).

At the end of these considerations of Dieste's buildings, it is important to note a general aspect of his architectural conceptualizations.

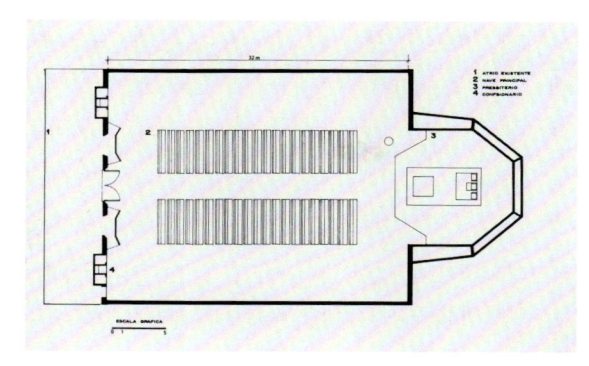

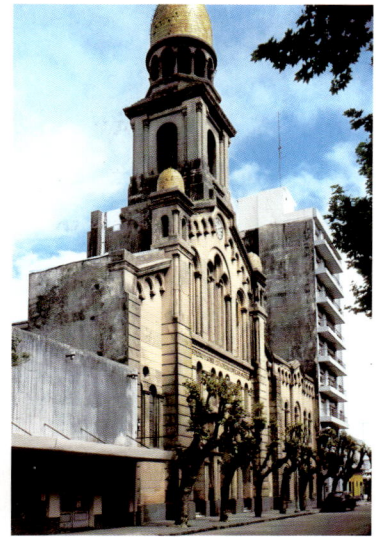

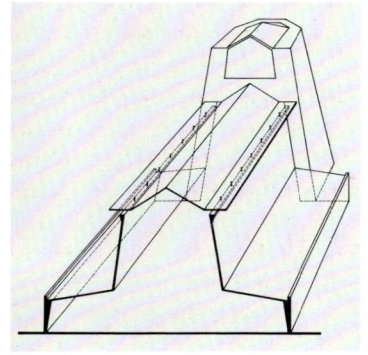

㉗ *(top left)*
Church of San Pedro, Durazno, Uruguay. Plan

㉘ *(top right)*
San Pedro, Durazno. Retained facade on the main square

㉙ *(center left)*
San Pedro, Durazno. Interior, angle view toward altar and side aisle

㉚ *(center right)*
San Pedro, Durazno. Axonometric section

㉛ *(bottom left)*
San Pedro, Durazno. Interior, rose window

With the exception of an uncompleted church in Montevideo, all Dieste's buildings—even the richly curved church at Atlántida—begin from a simple rectangular plan. If the plan of San Pedro (fig. 27) were all we knew of this church, we might easily find it a rather uninteresting, even dumb, conception. Here, as in all Dieste's buildings, it is not the plan, not the exteriors, but the sections, both transverse and longitudinal, that yield all the critical qualities of structure, space, and, perhaps still more importantly, light. Axonometric sections, as that for San Pedro (fig. 30), are key to the understanding of Dieste's work. One must emphasize that this observation is not about architectural graphics, but rather about a fundamentally important conception of architecture.

Before reaching some concluding remarks, one should note a few other characteristic types of structure by Dieste. Among his towers are the campanile at the church in Atlántida, a 60-meters-tall television tower in Maldonado, and numerous water towers, including a rather small one at the resort town of Las Vegas, near Atlántida (27 meters tall; 31,700 gallon capacity; 1966; fig. 32). The perforated shaft occurs in many of these towers and was a matter of careful consideration by Dieste. The perforations reduced wind loads and admitted light to the access stairs or ladders. Perhaps most important was the matter of construction: planks could be placed across the diameter of the shaft, from one opening to another, thus providing an efficient self-scaffolding. But visual matters were of similar importance. The perforations yielded vertical piers that emphasized the upward thrust of the shaft.

Water tower, Las Vegas, Uruguay

Dieste weighed the effects of having horizontal rings of perforations versus the diagonally displaced ones that he used. Visually he recognized that his solution enhanced, again, the vertical thrust of the shaft as opposed to an apparent layering from rings of openings. Further, since the shafts are tapered as they rise, the circumference changes. With the staggered perforations, the changing dimension can be taken up in the openings, thus preserving piers of constant dimensions without cut bricks. The towers again demonstrate Dieste's attention to architectural nuance while solving his problems of structure and construction.

Dieste built several horizontal silos, the last of which, and the largest, was that for the Navios company in Nueva Palmira (77,500 square feet; 1996–97; fig. 33). Light is a negative for a grain silo, so this is a continuous double-curvature vault. The site development of the project is almost that of an art installation. With a span of 45 meters, the scale is vast and the effect commensurate (fig. 34).

The poet Rafael Dieste (1899–1981) was an uncle of Eladio. One of his poems speaks of a miller building his mill and engaging in his trade. The poet observes that all the miller's work was directed to practical purposes, down-to-earth and necessary ones, but he also finds deep meaning in his work. The horizontal silo is one of the more humble kinds of structures Dieste built, a volume just to hold wheat. But, in the end, here too there is poetry, something of the sublime, in the structure and in the wheat.

An image of an earlier silo facilitates an understanding of Dieste's construction techniques (1975–78; fig. 35). A huge movable formwork, shown here as it has been lowered and is being moved to the next position, allows construction of one unit of a double-curvature vault (30-meter span). The timber surface of the formwork has a grid of little sticks that allows unskilled laborers to position the bricks. Small reinforcing bars are laid between the bricks, and the joints are mortared. Depending on the design, additional reinforcing is laid over the bricks before the entire surface receives a cement parging that is simultaneously structural and weather-resistant.

One of Dieste's sons, the structural engineer Antonio Dieste, has raised questions about the viability of his father's building program in today's economy, even in Uruguay. And, in North America or Europe, it seems even less viable.[1] He also asked: What innovations are necessary to preserve and advance reinforced ceramics? Such questions are ones that his father would have recognized. Today, the firm of Dieste y Montañez continues under the direction of another son, Eduardo Dieste, and with a long-time, excellent engineer-collaborator, Gonzalo Larrambebere. As we speak (2005), Dieste y Montañez is building another of the vast silos at Nueva Palmira.

So Dieste construction continues, but it may be well to consider whether this building program, with its simplicities and efficiencies, may not hold special promise for those lands with severely diminished resources. The contemporary, noted Finnish architectural firm of Heikkinen and Komonen, architects of the high-tech and elegant Finnish Embassy in Washington, D. C. (1990–94), has also taken the responsibility to aid in construction of necessary social facilities in the

33 Navios horizontal silo no. 4, Nueva Palmira, Uruguay

34 Navios silo, Nueva Palmira. Interior

35
CADYL horizontal silo, Young, Uruguay. Construction photo

African nation of Guinea.[2] They did not employ unusual structural systems, but together with their collaborators they trained local people in the production of ceramic elements of several kinds and designed model buildings in masonry and limited use of wood. There is every reason to think that Dieste's techniques, using wood only in formwork, could address an even wider range of problems effectively. One example known to me is the water tower constructed by one of the communities in Auroville, India.

In reaching a few concluding remarks, I ask the reader to keep in mind the image of Eladio Dieste (fig. 2) and of his rose window in San Pedro (fig. 31). I want to comment further on Dieste's strong traits of rationality, architectural acuity, and social responsibility, but I would also like you to bear in mind the twinkle in his eyes that is mirrored in the tour de force of the rose window. Dieste's face reveals both intelligence and charm, suggesting that he could accomplish something very serious and at the same time entertain pure inventions for the sheer joy of it. In San Pedro there is real substance to his thinking in the folded plate construction and the entire conception of the space, but also a playfulness and wit.

Dieste was a builder who worked from first principles, innovating with traditional materials and providing a model that might be applied in other ways. The significance of Dieste does not end there. He was born, lived, and dominantly built in Uruguay, a small country of high educational and cultural level, but not of generous resources. His choice of materials and still more the economy, not only financial economy, of his works systematically addressed the conditions of his country and potentially those of many under-resourced countries in an era of superpowers and globalization.

Dieste knew full well that ample resources, wealth, and power were not enough to assure a sound environment or humane society. Contending with constraints may deny many material advantages, but not the opportunity to think and work contributively—still less the opportunity to build a fulfilling social state. Dieste appreciated the simple farmers and workers of his country. The church in Atlántada is essentially for and dedicated to them.

Dieste was also a man of culture. He loved classical music. His care for literature is witnessed in his affection for his uncle, the poet Rafael Dieste. He moved in the avant-garde circle that developed around the modernist painter, Joaquin Torres-Garcia (1874–1949), who became a Dieste family friend. Dieste wrote about his engineering innovations and accomplishments, of course. Yet those accomplishments lay at the core of ever-expanding concerns. He saw the negative impact of much of modern development on cityscapes and the countryside. He understood that art and architecture were integral to the making of the admired buildings and cities of the past and had to be part of any desirable future.

Dieste was a man of religion. This is evidenced in the sheer architectural, and one can only say, spiritual quality of his churches. The way in which his churches intentionally unite the worshipers with the clergy is evidence of his embrace of the liberalizing forces in the church. But religion makes little appearance in his writings. His concerns were for social justice, for the opportunities of the less-resourced peoples and countries, for thought and action through what he termed "cosmic economy." His life, his thought, and his work were integral and integrated.

Notes

1   Antonio Dieste, "A Prospect for Structural Ceramics," in Stanford Anderson, ed., *Eladio Dieste, Innovation in Structural Art* (New York: Princeton Architectural Press, 2004), 220–22.
2   Mikko Heikkinen and Markku Komonen, *Before Next: Learning the Roots* (Helsinki: Museum of Finnish Architecture, 2002).

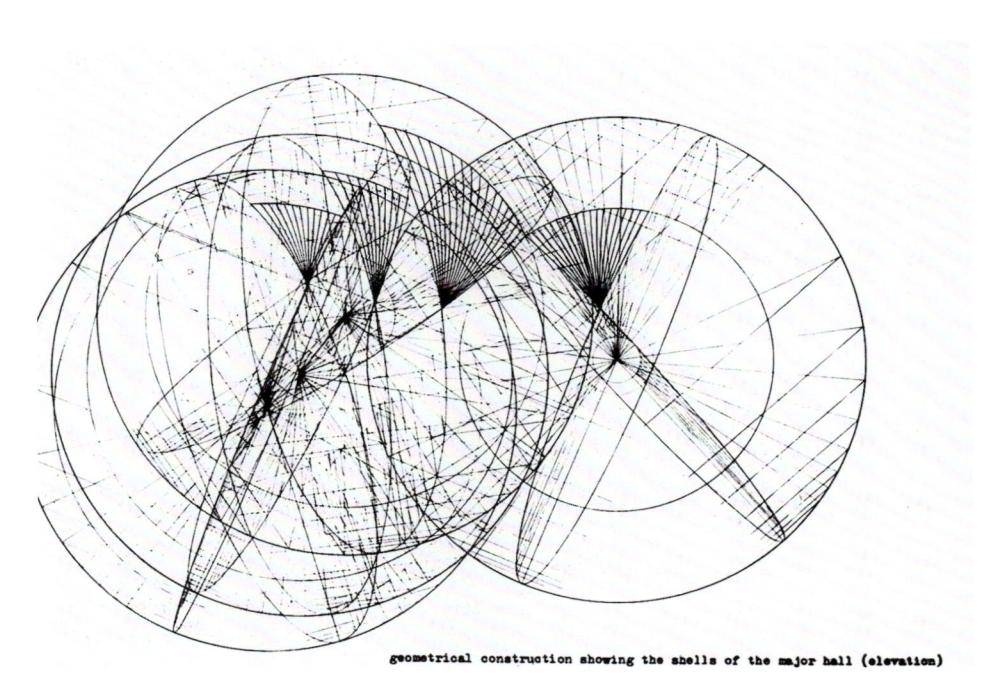

geometrical construction showing the shells of the major hall (elevation)

❶ and ❷
Jørn Utzon and Hall, Todd, and Littleton with
Ove Arup & Partners. Sydney Opera House,
Sydney, Australia

48 | THE FELIX CANDELA LECTURES

Informal Networks

CECIL BALMOND

I've called this talk "Informal Networks" because it expresses my approach to structural engineering. "Informal" because I do not necessarily accept a fixed symmetry for a structural frame—I allow notions to slip, jump, overlap, and trace to enter the dialogue as dynamic elements—and "Networks" to convey the idea of structure as a connective part through pattern. This allows me to develop a mobile sense of geometry and let structure be a catalyst for architecture. This lecture is a kind of look back on my career, a summing up of where I am now. When you look back, you find that there are all sorts of things that have influenced you—your culture, your research, and so on—and each element also has an impact on the others.

I was born and raised in Sri Lanka. It is a wonderful, very lush country. It is also an ancient country with a civilization more than 2,500 years old. I was brought up in a multicultural environment, and I spoke both English and Sinhalese. I was a Christian who had Buddhist, Hindu, and Muslim friends. I therefore grew up with a certain complexity of images that I guess was buried somewhere in my psyche and which seems to have emerged in my later work.

I also encountered early on a thing called "magic," a kind of mystery of the irrational. People walk on fire. If you and I tried to do it, we would be severely burned, but their faith allowed them to walk, and that is a deep mystery. Similarly, there were people skewering themselves with metal objects well before there was a fashion for body armor in the West. As a child, I would open the door and find a man, pierced with hooks and spikes, who would scare the hell out of me!

I grew up with these images and then went on to a rational study of physics, mathematics, and engineering. After graduating from the University of Ceylon, I completed my studies in England. I found that engineering had images with beautiful stress patterns; there was a certain wonder to them, and I was very happily engaged with this. When I started practicing as an engineer, however, I found that most of what I was doing was framing things up and being very proud of ratios of length to span. I quickly learned the methods for steel and concrete. Architecture at that time, I thought, seemed to be concerned with

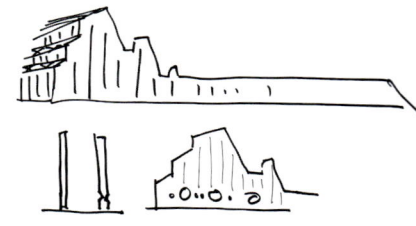

3 and 4
Carlsberg Brewery, Northampton, UK

external placement, the aesthetic placement of objects in space. Very little effort was given to the interior structure of the design, which was left to the creativity of the engineer. I saw that engineering did have a certain power to influence, as evidenced by such projects as the Sydney Opera House (1973; figs. 1, 2) or the Penguin Pool in the London Zoo, in Regent's Park (1934), where the engineer was Ove Arup.

I was very fortunate that the Carlsberg Brewery in Northampton, UK, was my first major project (1973; figs. 3, 4). It was a seminal experience in my formative years as an engineer, as I was beginning to move toward architecture. Essentially, the building was composed of a number of huge concrete walls that flanked the machinery. Then there was a glass facade that spanned 30 meters vertically, with two offsets in elevation between the ground and the roof; the in-plane stiffness brought the forces back to the walls. The trusses were designed like a bridge beam, with bearings at either end to let it move because I was terrified that the glass would crack: it covered an immense 10,000 square-foot opening!

In the brew house itself, we wanted to put the equipment on two levels. This plan encountered significant resistance from the Carlsberg clients, but eventually we convinced them that it could be done. Both the architect, Knud Munk, and I aimed for an almost theatrical installation. In contrast to an industrial installation, in which there would be bracing everywhere, I found that I could actually make the system work without any bracing at all, based on the sizes already needed to prevent sway, and with the additional use of good friction-grip bolting. I checked the vibrations thoroughly and they were fine.

However, one worrying moment occurred, which remains in my mind. There was a particular wall that was about 33 meters high at its highest point but 1 meter wide, and which had to be freestanding for some time before the roof was connected. I discovered at some point into the project that I had forgotten that the service engineers needed a series of major holes punched through this wall to get the brewing pipes out! After some deliberation, I eventually made this work by more precise calculations, allowing for lower factors of safety in the temporary case.

The entire project was 220 meters long, including the bottling plant, the fermenting tanks, and the brew house. The roof was a folded plate in concrete, which I did by hand (there were no fast computers in those days) I am very proud to say! Timber shutters attached to steel trusses stiffened this roof so that, when the concrete was pumped in, the wet load was supported by the shutters and the skeletal truss inside. Then the concrete set and the whole thing worked very well, picking up any residual stresses.

I learned a good deal on this job, especially from a technical point of view. Most importantly, during this period Ove Arup himself was reviewing my work once every month. The critique sessions that I had with him were marvelous. Not only did he evaluate my work as an engineer but also as a designer, and the experience was very informative for me. I am glad to say that, after twenty-four years, the project won another prize last year for a most enduring building (the Brewery also won many prizes at the time of its completion). It remains a sort of cathedral to brewing. Carlsberg was also a great client to work for—lots of beer and free cigars. In fact, I do not remember the end of any of the site meetings!

In the early 1980s I worked with James Stirling on the Neue Staatsgalerie, Stuttgart (figs. 5, 6), with its now-famous public walkway: people are taken from one street up through the walkway and out, without ever

5 and **6**
Neue Staatsgalerie, Stuttgart, Germany

entering the gallery. I mention this project because, apart from all the engineering (and there were many elements going on there), it was the first job in which I took on issues that were wholly architectural. I found myself arguing for spatial qualities from an architectural point of view while, of course, using structural know-how to validate my position.

The changing exhibition space and the lecture spaces had no columns in them. Stirling had kept this area free. It seems like a very small thing now, but I argued to put in columns. This was not from a pragmatic point of view, to make my life simpler by inserting the columns, but because I felt that the space needed it, and it made more sense in the context of what was above. After some discussions, Jim agreed, and that was a huge moment for me as a young engineer: that a great architect was listening and that I could influence an architectural thesis. With his panache, Jim made the tops of the columns into wonderful drums, so they became what I call "ice cream cone columns."

My older colleagues at the time advised me to insert expansion joints. The plan was about 110 meters lengthwise and 100 meters across. There were large, exposed areas that would heat, expand, and later contract, and which needed control. Everywhere I tried to put an expansion joint, however, I felt from an aesthetic point of view that it was not helping the architecture. So I decided against an expansion joint. I found that, in terms of the cost of reinforcement going in to

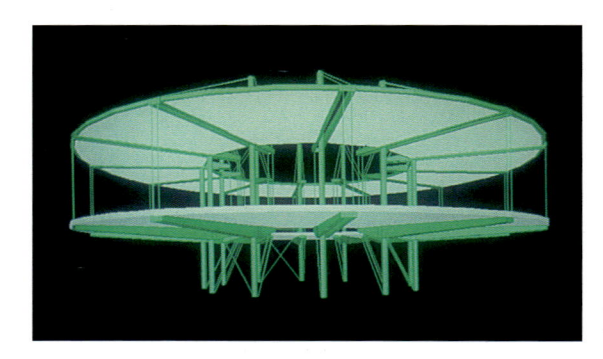

❼ and ❽
Temporary exhibition stand, Dutch Telecom

52 | THE FELIX CANDELA LECTURES

9
House, Bordeaux,
France

resist some of the high tensions or forces at certain corners, the effect was very small. I learned a lot from this project as well, about how not necessarily to take structural training and traditions at face value but to work something out on first principles.

I worked with Jim on lots of projects. In the United States we did the Science Library at Irvine, California. We finished the last job before Jim died in 1992. It was the Polytechnic in Singapore, which had strong geometric forms. In London we undertook an office building on a site where Ludwig Mies van der Rohe had tried before—the address was No. 1 Poultry, right by the Bank of England in the City of London (1998). It is a building with a strong triangular form and an embedded circle. Jim was a geometrician, albeit a very special kind of one. I read that there was a time when he never built anything, but I knew Stirling had a particular capacity to build in his head. I learned a lot about three-dimensional thinking from him.

In the mid-1980s I ventured into some high-tech work, which included a temporary exhibition stand for Dutch Telecom (figs. 7, 8). The company mounted the stand, then demounted it and took it around The Netherlands. It won a steel prize, which was quite easy to do as there was not much steel work being done. The Dutch mainly built in concrete at that time. Still, I was happy to get a prize! But high-tech work did not really inspire me. Innately, instinctively, I preferred projects like Felix Candela's: the fluid motion and precision—no edge beams, nothing, giving a kind of "Look, No Hands!" effect—and the elegance. I also liked Le Corbusier's Ronchamp. Pier Luigi Nervi was a big hero to all the engineers of the time, but his work did not particularly excite me. I could admire the precision, but the idea of an evenly distributing structure, though it had its moments, was a bit too clinical for me. I preferred something more varied, in which the typology was mixed: a shear wall, a truss feature, a dome that would be truncated, for instance. Or something like the house in Bordeaux that Rem Koolhaas and I completed five or six years ago: a big concrete culvert sort of box held on eccentric supports (1998; fig. 9).

BALMOND | 53

I have had a very significant collaboration with Koolhaas beginning in 1986. A lot has been written about Rem but something that is probably not known—and why we have got on so well and established a good working relationship—is that he was interested from the outset in exploring structure. For me, the work with Koolhaas has been a kind of intense exploration of structure as episode. It is not structure in the uniform sense; it is not structure that distributes itself everywhere; it is structure that works in local conditions and does specific things.

So for the Kunsthal, the art gallery in Rotterdam (1992), instead of a traditional brace that was going to be hidden in a roof plane, we exposed it and suggested a curved arch. It became an architectural device and a kind of motif in the space. And then for the Grand Palais in Lille (1994), the inclined bracing is seen within the exhibition space. We could have hidden it on top and taken it out to the sides but Rem said, "Why not bring it down so we celebrate the brace itself?" For the roof, I proposed a hybrid timber/steel solution, which was, in the end, quite inexpensive to do: structural steel on the top and reinforcement bars bent up and welded to a T. The assembly came to the site and was bolted to a plank, serving as a tension cord, a sort of do-it-yourself beam. Instead of the much harsher industrial roof that this could have been, the timber gave a very soft quality. So again, hybrid conditions in material terms were something I became interested in. Whatever could condescendingly be called impure interested me.

In the library in Jussieu, Paris—which we did not get to build in the end because of a change in the French government—there was a spiraling form. I used the entire floor sweep as one massive brace so that there were no discrete diagonals. The floor itself did the job.

Something that I have been exploring consistently with Rem over the years is the idea of using a minimum structure initially and, then, if the element gets into trouble, and only then, do we seek to take remedial action. This may assume the form of a prop, doubling up, or whatever else is deemed suitable. It is an ad-hoc procedure that is not smoothly ironed out but that we feel gives a great dynamic to the space.

The Seattle Public Library project (2004; fig. 10) was designed to have three self-contained boxes. A minimum amount of vertical support was put in place—only about eleven or twelve columns—indeed, very minimal for the size of the space. The tectonic boxes were supported in this minimum sense, with a thin wrap around them for seismicity in case of an earthquake. Where the skin needed more strength, I doubled up the structure to beautiful effect; when gravity needed a bit of help, props came in on an ad-hoc basis. This gave it an immense power, a kind of continuous reading to the space, and a surprise as you turned the corners. I think this will remain one of Koolhaas's best buildings.

Besides using only a minimum structure initially, the other thing I introduced Rem to was the Vierendeel girder. Here was something not used much in architecture but rather in railway design from about 1880 to 1900. This girder is story height, thus enabling you to have a

10 Seattle Public Library, Seattle, Washington

11 Casa da Musica, Porto, Portugal. Structural elements

multistory building with no columns at all on alternate stories. We used these for the ZKM project. Koolhaas was fascinated with trying to compile spaces differently, using elements of structure but in a wholly architectural way. Witness the Concert Hall in Porto (fig. 11).

For the CCTV project in Beijing, we designed the bracing to go around the structure in a changing pattern (fig. 12). The idea of a classical core was completely lost. This new kind of configuration put the onus on the skin throughout, making it impossible to second-guess what was going on. The structure was 260 meters high (just under 800 feet)—comprising 5 million square feet—a monster of a building! Where stress was needed, where there was more intensity, the pattern doubled-up so there was a double rhythm. If you took that pattern and then removed the columns and the floors, a very unusual pattern emerged (figs. 13, 14). By forcing this onto the facade, the effect was to give an open-ended and changing look.

The seismic aspect of the structure was also interesting; the whole looped building vibrated approximately 4.1 seconds. This was the first mode as it cantilevered out and moved over into the space. The second mode was when it moved as a portal, and it was quite stable in that direction. Then it moved into the torsion mode. This was the most interesting phase for me (and also the one that caused me some concern), but here I also got a good result, 2.3 seconds. I felt that this was about the right ratio; if it had crept up toward 3 seconds, I would have been worried. I would not have wanted the torsion mode getting mixed

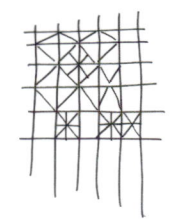

12 *(left)*
CCTV Headquarters, Beijing

13 and **14** *(above)*
Structural patterns for CCTV Headquarters

up with the main fundamental mode, which was not a good place to be. Rem and I built huge models to fully inculcate our teams with the sheer size of the building. It was bigger than anything we had ever done—the plate sizes, the steel, the columns, everything.

The project took nine months to obtain all approvals, with twelve Chinese professors lined up against us, each an expert in his own field: a hugely daunting experience! However, we worked out our strategy very carefully, like a military campaign, and I am glad to say that last month we received the full building approval with no qualifications at all. Anyone who has had experience working in China will understand what a feat this was! Additionally, we are within the time frame to get this project built by 2009.

Getting back to what has influenced me, I would say that music has been and continues to be a major part of my life. I listen to it all the time and, if no one is around, I even play it a little. I study it extensively, particularly Bach, whose work amazes me. Listen to the Partita for Violin in D minor, called the "Chaconne," for example. This piece is very tonal, very harmonic, and yet quite serial. Every four bars there is a repeating harmonic pattern. It is like a hidden algorithm at work, quite amazing. It is pure genius how Bach did this.

I am also interested in dance: the idea of movements embedded one into the other—a kind of dynamic symmetry—and the strict geometry of it. If you saw an Indian dancer, for instance, you would see his or her hands moving freely and you might think that this was an improvised free-form movement, but it is actually performed to a set of very rigidly controlled rules. There is a hidden geometry at work, although you do not need to know about it: you merely enjoy the performance.

I am also fascinated with folk art. I do not study it exactly but just look at it: aboriginal art, the art of Sri Lanka, of the Andaman Islands, any folk art. It often looks symmetrical at first glance but actually it is not. Each part is different, changing, not compartmented or static, and each is layered over the other.

And then there is the newer mathematical art from the late 1970s and early 1980s: the famous fractals of Mandelbrot. The pattern keeps repeating, and if you zoom into any one of its parts you find the same motif repeating again. It is scaleless because it goes on and on and on, and that is the sign of a fractal: that at any one level you get all scales embedded in the same picture. So it is amazing, and it does not have to be chaotic. You can get a symmetrical fractal.

Take our own DNA molecule. What do you see if you look down the molecule? Imagine the four phosphate protein base, called A, C, G, and T. When they come together, they form a spine and start joining up. They have no idea that they are going to make a DNA molecule. What actually happens is that you get a little rule that works: whenever you get an A, it can only connect to a G, and if you get a C then it can only connect to a T. ACGT stitches together in such a way as to create the DNA form, and all the mutations and variations, and thereafter life.

This idea of form emerging is a thing that really fascinates me: from a random start, how does it happen? It happens through control and continual feedback. Take the knight or horse move in chess. It is a two-one random move. It looks very arbitrary, but wherever you start on the chessboard, if you play it out, you get these patterns that emerge, a sort of equilibrium. If you start somewhere else on the chessboard, you get another pattern, and so it goes.

This idea of an answer, a solution, a structure, an organization evolving, is something that I have been looking into recently. What happens if I start at the half-point and draw a line to the third point? What if I do it again on the next side: half to third, half to third? Then I finish this square off and, when I have closed it, I do it again: half to third. If I keep doing this, I begin to get a complex spiraling web (fig. 15). If I extend all the lines, I have the chance for a piece of architecture. Now imagine this is just a set of lines but imagine each line becomes a 15-inch steel plate of ½-inch thickness. Imagine also that those particular lines, the ones that are the principal structural lines, are 1½ inches thick. Now cut out the corners and you get a pavilion!

This idea was developed into the 2002 Pavilion for the Serpentine Gallery in collaboration with the architect Toyo Ito. Ito and I decided to checkerboard the pattern (figs. 16–19). We put aluminium panels in, painted them white, alternating aluminum with glass. It was a great success. We brought some of the lines across the floor and also brought the ground in with the grass. If you followed the algorithm strictly, as the line comes down it can avoid having a corner touch the ground. Of course, the corner is not important because the structure has just whizzed across the diagonals and the corner could lift off, as it were.

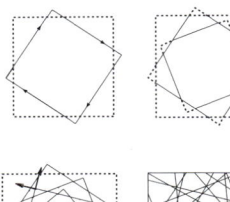

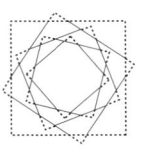

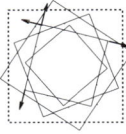

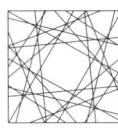

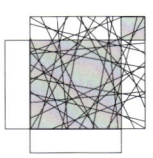

15 *(above)*
Drawings of spiraling webs, Serpentine Gallery, London

16 *(left)*
Pavilion, Serpentine Gallery, Hyde Park, London

17 – **19** *(below)*
Checkerboard pattern for pavilion

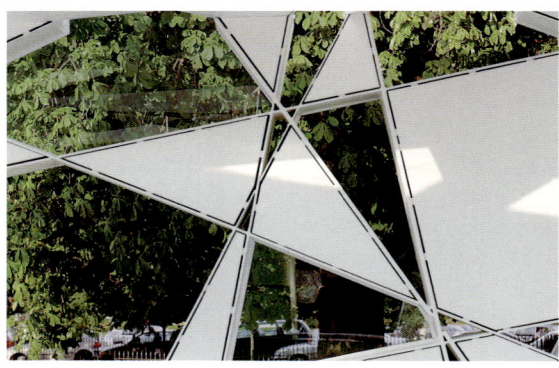

58 | THE FELIX CANDELA LECTURES

I also plotted the pattern, extending it to, say, three quarters of a mile in each direction. A little black dot represented the actual pavilion, and the ultimate symmetries could be seen through its spiraling, dynamic form. If you zoomed in, you got the impression of a kind of interconnected world. It would have been unthinkable to make this ten years ago but with computer fabrication we went straight from the design to Xsteel—special project software that cuts the steel—and straight from the computer modeler to the factory and out. Everything was a one-off, nothing was repeated. The total weight came to fifty tons (I checked the equivalent grillage, which came to about fifty-two tons). You had to appreciate the amazing aspects of a pattern that had been compiled and that had both an open and a closed nature. You could not second-guess it. It was a kind of meditation on geometry itself.

I worked with Daniel Libeskind on an art gallery for the Victoria and Albert Museum in London (2003). There were planes of wall about 12 meters tall and six giant turns of the structure. Internally it was comprised of about eight floors, plus a restaurant on top. Inside there was an amazing sense of space and of a rotation going on in that space. Every two floors, the edge was cut out of the floor so that you could see the wall planes moving. There was a slow sense of a displaced vertical. If you looked down, you saw a spiraling form. In nature we have the logarithmic spiral—the snail, the tornado, and so on—and we have the Archimedean spiral, which is similar to a watch spring that winds up regularly, gradually increasing. These are both fixed-centered spirals. If that fixed center itself moves, you get what I call a chaotic spiral since the center is moving in a way that is relative to what is going on. (The algorithm is a simple one, and one that I have described in my book, *Informal*.) Using this spiral as an original line, I am able to pull it up into a third dimension and flesh it out with structure, the walls bearing on each other to support the load. The original blue line of the spiral on our first plot became a fully manifested form, ultimately going through the center of the 12-to-18-inch-thick concrete walls of the structure (fig. 20).

As interesting as the form was, there was something much more intriguing in the project for me, which had to do with tiling. Libeskind and I both knew we could not use regular tiling: it could not be some kind of repeat tiling. It had to be something different, what I call a mathematic mosaic, something that moves around. I found these three tiles, which I refer to as P, Q, and R (fig. 21). If you put them together, just those three shapes, you can tile an infinite plane (figs. 22, 23). It looks similar in places but actually nowhere does the pattern repeat. Just when you think you have worked it out, you find an interruption. It is never the same—forever! This tiling was invented by an American mathematician, Robert Ammann.

The secret to this tiling and all such tiling, which is called aperiodic, is this hidden construction. A hidden grid goes through all the vertices of the tiling (fig. 25). The rectangles are Golden Rectangles, which were used in classical art and architecture. There is one other aspect of the tile that is amazing. Each tile breaks down into four smaller Ps, one R and one Q. So every tile has its own shape, a smaller part embedded in it and the other two tiles (fig. 26).

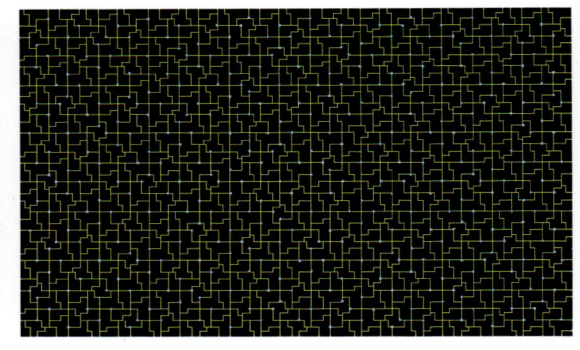

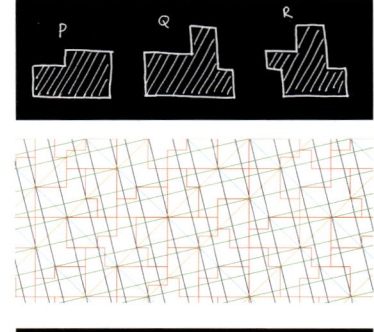

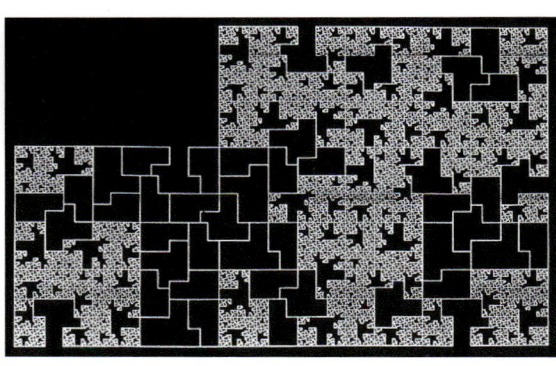

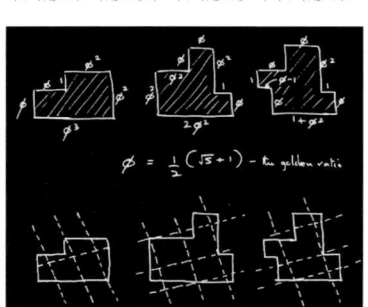

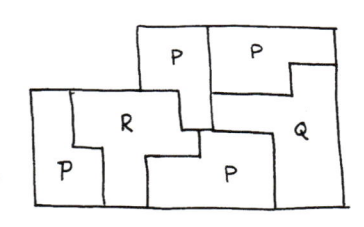

I thought that if we used just the Ammann tiling, it would not provide the full dynamic that we wanted for the project. So I decided, with my team, to take it further. If you imagine using a cut-and-paste program on a computer, you take the pattern, P, and put it all back successively, but every time you do it you do not degenerate R. You keep the shape blocked. Here is the first R stopped, here is the next, here is the next, and so on, until you get a fractal (fig. 24). It actually is a branching that happens, like a tree: you get finer and finer branches until you get to the leaves and the buds. This branching tends to give a pattern that

20 *(top left)*
Victoria and Albert Museum, London. Model

21 and **22** *(center left)*
Drawings of tile shapes and computer drawing of tiles

23 *(top right)*
Computer drawing of tiles

24 *(center right)*
Fractal of tiles pattern

25 *(bottom left)*
Drawings of tile mathematics

26 *(bottom right)*
Diagram of tile shapes, embedded

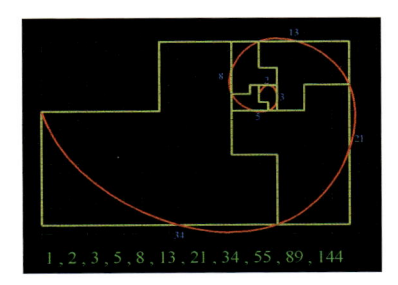

㉗
Victoria and Albert Museum, London

㉘
Schematic drawing with Fibonacci numbers

has a density on two particular lines. What you have done is take your first real zoom into infinity. If you have never been there, this is what it is like! These are not white lines. They are infinitely dense, with black patterns, which is what is amazing about them!

We decided to use off-white tiles, so they were all one color. Libeskind embossed part of the pattern. In this way we got a very interesting reading on the form (fig. 27). It was not tiling that you are used to: it was almost like text being read on the form, to very powerful effect.

The tiling pattern can also be derived from the Fibonacci numbers, which are found in nature (fig. 28). It is quite fascinating how numbers themselves give form. A simple 1 and 1 right-angle triangle gives $\sqrt{2}$; if you add another 1, you get $\sqrt{3}$. This is how people found square roots in ancient times. It enables you to get a spiral out of it.

BALMOND | 61

I used numbers in a very simple way, first with Alvaro Siza. It was a simple rule of numbers 0, 1, and -1 that just shifted up and down, this way and that, and went to form the roof for the Portuguese Pavilion in Hannover, Germany (2000). Siza and I sat there trying to draw waves, but a free form cannot be drawn. It will always be clumsy. It is actually better to go into an algorithm or some rule by which you can create the form. That is what we found. There was more freedom in the rule (fig. 29).

I worked with Toyo Ito on a project where I used numbers to generate form. Instead of straight columns, I displaced the columns by a simple vector rule; we called the answer "Dancing Columns" (fig. 30). If you looked down the column movement, it was like a code at work, and repeating patterns were formed. What was originally random was shaking down to give some kind of order. If it was to be a building, for a department store, say, you could work out that the slightly artificial slopes may be beneficial for your layout. There was also the facade, coming straight from the numbers. No hand had drawn anything here nor had we any idea what it looked like; this is what you got simply from the numbers.

The shapes also inverted themselves and gave rise to a set number of shapes. If the facade was to be precast, then only six types of panel were needed (fig. 31). We found an algorithm for creating the pattern (fig. 32), and it was like the DNA strand of the four bases with sequential relations. Provided you follow this rule, you can keep going on and on, never the same, always changing but perfectly pragmatic, and you can precast it and there are only six panel types.

Work of this type is being done in London at the moment with a group that I have called the Advanced Geometry Unit: a collection of architects and engineers; a quantum physicist, who brings in the advanced mathematics; and a game theorist. The unit has grown now to about twenty-five people. I am hoping to extend this practice to the

29 *(top left)*
Portuguese Pavilion, Hannover, Germany

30 *(bottom left)*
Drawing of column profiles

31 *(top right)*
Repetition of panels

32 *(bottom right)*
Algorithm for generating panels

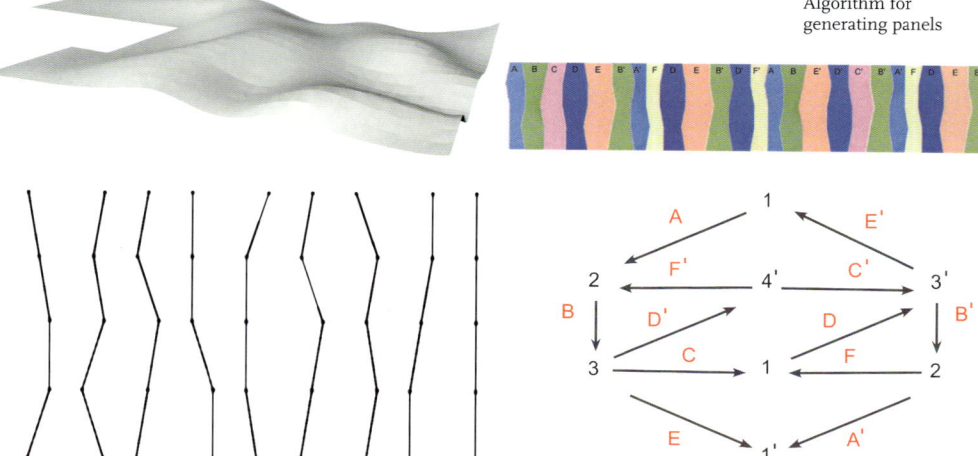

62 | THE FELIX CANDELA LECTURES

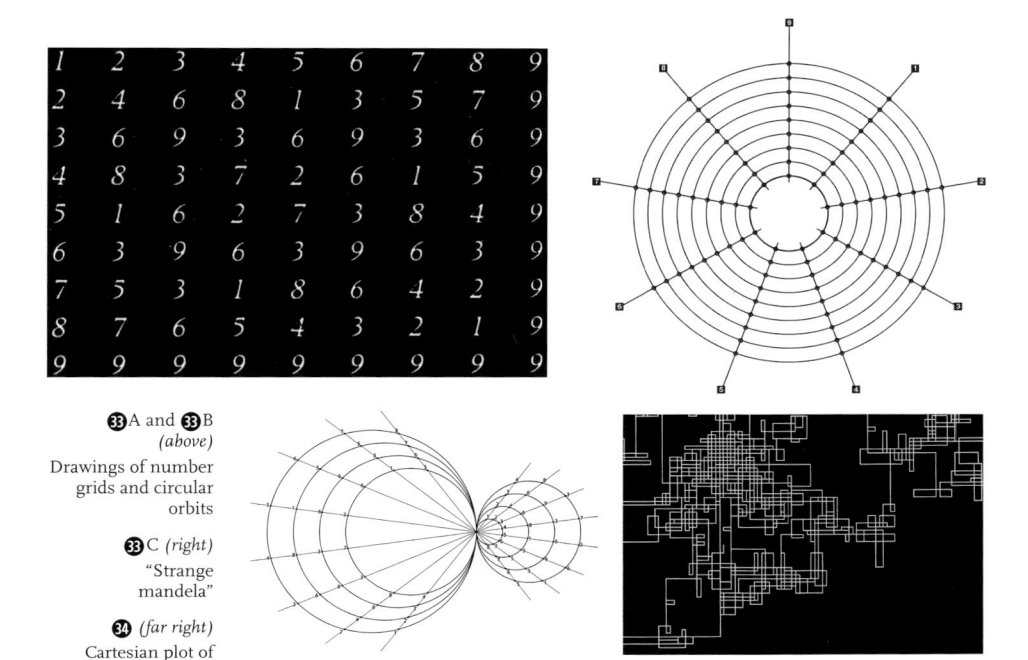

1	2	3	4	5	6	7	8	9
2	4	6	8	1	3	5	7	9
3	6	9	3	6	9	3	6	9
4	8	3	7	2	6	1	5	9
5	1	6	2	7	3	8	4	9
6	3	9	6	3	9	6	3	9
7	5	3	1	8	6	4	2	9
8	7	6	5	4	3	2	1	9
9	9	9	9	9	9	9	9	9

33 A and **33** B *(above)*
Drawings of number grids and circular orbits

33 C *(right)*
"Strange mandela"

34 *(far right)*
Cartesian plot of number sequence

Arup offices in Tokyo and New York to develop this kind of advanced thinking in form making.

Another line of research has to do with number arrays. In my book, *Number 9*, I reduced the products of multiplication to rows of single digits (fig. 33A). To my surprise I found four patterns—the one-times, two-times, three-times, four-times tables—which get reversed. The two-times table is a reverse of the seven, one is the reverse of eight. This was arranged in a series of circular orbits (fig. 33B). I also found another kind of organization, a more contemporary one, which acts as a single point attractor; it is a kind of perpetual motion machine (fig. 33C). I experiment with structure from these patterns.

The two-times table can lead to a kind of molecular structure, which I started making more prismatic for a high-rise tower in China and where I found how strong this kind of weaving form was. If I use the three-times table, I get the more familiar straight-line-generating Antonio Gaudí forms. I can also get weirder shapes by applying the whole field of numbers. The one to nine values can also be graded as temperature or can be graded as light, and so a piece of work that you make as structure can also be informed by light and sound. In fact, that is the kind of thing that I have been doing recently in my teaching.

And if the numbers get fed back into themselves, a very rigid, benzene-ring kind of structure appears. Again you get freeform folded structures. If ninety degrees is the key you turn, you get a kind of scatter and, if you zoom in, it is like a city grid. Order appears seemingly out of nowhere (fig. 34). The random move is suddenly shaking down and giving nuclei of order.

And, of course, it is possible to get structure. A random walk, two random walks, two edges, and a sweep through with a certain software give you minimum surfaces. They are all efficient surfaces but they are forms that you would never think of. I am studying certain crystals, new kinds of crystal formations that are coming from a three-dimensional array of our multiplication tables, which is amazing. This work will be published later to have a wider audience. It is research-based and part of the teaching that I have been doing at Yale and Harvard.

One of my current projects is in Battersea, London: a huge £1 billion project to convert an old power station. One side will house a Hyatt hotel, and there will be another hotel on the other side. There is also a theater, 800,000 square feet of office space, some housing, and so on. This project was planned a couple of years ago, and I was asked for an alternative idea to the original plan. There are three levels of car park underneath everything in a huge basement, which has the slow movement of a double helix, one level moving through the car park and gradually emerging and ending up as a plateau for the office building. The other strand emerges in a different place. In the center of this rotation there is a void, so we took the theater and put it in the void. You enter and slip down into the car park. The theater roof is a public space itself, so there is a serial connectivity again: a public space, which is local to each office and yet all of it is part of a landscape. The whole project is really about a traveling edge.

So out of the geometry comes a solution that is unpredictable and unusual, and has now been adopted for the master plan of this project. You never get a rectangular wall, you always get a wedge form, and the shapes are just sliding past you as they slip into the ground or come up (fig. 35). The whole idea is that this is a landscape. In the car park you get these beautiful vistas, with the power station as a focus. We are using a mica concrete to give a shot feeling to the concrete. There are beautiful spaces in front of the office buildings where you can see what is going on, and there is a triangular grid operating underneath it all.

At night the glass crystal handrail moves down, and it will be lit to get some dramatic effects. The whole thing is a twenty-four-hour center (fig. 36). The theater is a hexagon. To install it above ground costs a lot of money with the external skin, etc., but by burying it, by paying a little bit more for the raking slopes, etc., the net cost is the same.

Various architects have been appointed to design the two hotels, two office buildings, and retail center. I am designing one of the office buildings that I call The Twist. It refers to the form, a movement that goes from the vertical up and over into the roof, and vice versa on the other side. The twisted form also arches to allow entry into the scheme and, if you drive past, the entire building signals entry. So it is not any one entry anymore, as in the old scheme: the entire building and its adjacency signal entry.

The form is made in this way: it is a wrap that spirals one way, then spirals the other way, and then solidifies into a glazing. A strict geometry

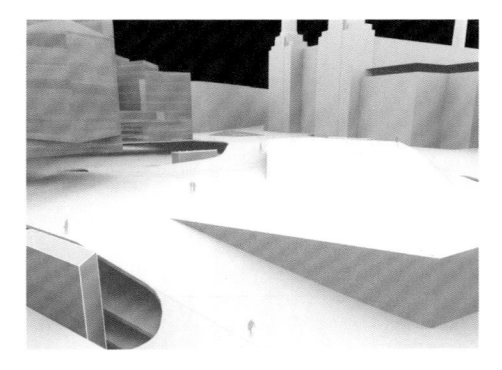

35 and 36
Master plan,
Battersea project,
London

is given, triangulating the whole thing because with a shape like this a normal structure would not be possible. There are no columns, no beams, as we understand them. This is a shell, a shell form that merges with a shear wall form, a deep truss, arches, and supports. The object is like a crystalline form. You could have glazing coming down from the wall of one side going to the underbelly. This creates a dramatic entrance.

We are now preparing the schematic design, and it is a very exciting project. The edges move, curving in space. There is a big internal span of about 50 meters, but the edges hold themselves up through this technique of spinning the beam on edge.

Another thought was, instead of crystalline form, what if we clad it with louvers? These louvers go up and, in turn, twist 180 or 360 degrees. Just as the beams twisted in plane in the space along the atrium, here they twist in elevation, so as you drive past, this building will open and close. I thought ultimately that it was a bit too industrial, and so we will probably revert to the crystalline form.

If I had to sum up what I have stated so far, I would say that I began my career inheriting the Cartesian world that we are sitting in, living in, believing in. But the more general, nonlinear and generic is actually what surrounds all of us—and the space we understand well—the Cartesian—is only one particular subset.

We have a choice now of two worlds. The first is the one we know: it is hierarchical; there is nothing wrong with it; there are a lot of inventions still to come; it is endless because human creativity knows no limit, so there is a lot here for us to do in architecture and engineering. But there is also another world: the more generic, nonlinear. To investigate this world forces architects and engineers to think of new ways to construct, using new materials—and ultimately new ways to think of architectural program. It will make architects and engineers work closer together. Once, there was no distinction between the architect and the engineer. They were the same, a building designer. What I see ahead is an intense, immediate exploration between architecture and engineering—and that is something to which I look forward very much.

❶
Alexander Graham Bell with his wife, who is holding the first space-frame structure, created by her husband.

A Life in Structural Engineering
LESLIE E. ROBERTSON

Overtly, this first image depicts Alexander Graham Bell bussing his wife (fig.1), but as we examine the photograph, we see that he is doing something that we all need to do extremely well: communicate. More specifically, he is communicating to his audience that he has designed the first space-frame structure and that his structure is very light, airy, and strong, as demonstrated by his wife's ability to lift it from the floor. Bell, a consummate innovator and communicator, used space frames for a variety of structures: kites, including powered kites, boats, a tower on his estate, and so forth. We may not all be great inventors, but each of us is able to learn to communicate our ideas to others, and to do so convincingly and with wit. Unfortunately, too many of us in the design professions fail in this most important pursuit.

IBM Building, Pittsburgh

The member sizes in this IBM Building (1963; fig. 2) were determined by making use of a technique that is, perhaps, one of our early signature approaches to structural engineering. The traditional approach had been to establish the loads and the yield point of the steel and then to determine the required member properties. Instead, we first determined member loads and member properties, all based on strength, performance, and functional and aesthetic requirements, and then determined the required yield point of the steel needed to meet those requirements. In the old way, the yield point was a given; in our approach, the yield point was the answer. Structural analysis for this IBM Building was by an iterative technique, using desk calculators.

❷ IBM Building, Pittsburgh

With this approach, we found the need for steels ranging from a low of 36ksi (250MPa), painted yellow in the photograph (fig. 3), to a high of 100ksi (690MPa), painted red, all while maintaining a structure with the largest members being a pair of 150-millimeter angles. This was our first really big step in the prefabrication of large structural steel panels (fig. 4). The architects were Curtis and Davis of New Orleans.

❸ IBM Building. Painted steel strengths on actual building during construction

❹ IBM Building. Prefabricated steel panels

World Trade Center, New York

Minoru Yamasaki was totally responsible for our firm's establishing an office in New York, with its initial project, the design of the World Trade Center (fig. 5). Yamasaki was a fascinating and seemingly complex man, designing almost exclusively as a critic would. That is, nearly all elements were examined in models; he would look at them and say, "Well, let's change this and let's make that more slender," and so forth. Because he was much older than I —I was then just thirty-five— it took a lot of getting used to; still, we became good friends. How many architects would have the courage to entrust the design of a World Trade Center to a youngster, not yet thirty-five?

Each floor of the World Trade Center, completed in 1970, was almost an acre in size, with clear spans of 10.6 to 18.3 meters. Never before or since has prefabrication been used to this extent. Floor panels were as large as 6.1 meters wide by 18.3 meters long, complete with a metal deck, electrical distribution cells, and so forth; column/spandrel wall panels were constructed three columns wide and three stories high (fig. 6).

The structural concept was that of a "tube," with all of the lateral forces from the wind or an earthquake taken in the column/spandrel system of the outside wall. This system *required* that the area between the services core and the outside wall be free of columns, a circumstance that pleased both the architects and the rental agents (fig. 7).

There has been much discussion of the identity of the engineer who first conceived of the concept of the "tube," with a handful of engineers being so named. It seems to me that it was an idea in need of realization, making it not unlikely that several engineers more or less simultaneously came up with the same concept. In any event, I neither lay claim to nor do I disavow authorship, believing instead that ideas are creatures of their time, not of an individual.

Analyses for the Center were accomplished on an IBM 1620 computer, the most powerful medium-scale solid-state computer of its time. Each of our external disk drives was the size of a washing machine. The multiplexer was the size of a large refrigerator. Input was by IBM punched cards read onto the disks. Output was by line printer and by IBM punched cards. While our experience was limited to but two twenty-story buildings, other engineers, who had long been established in New York, had designed scores of high-rise buildings. Accordingly, we set out to find what was really good and not so good about the work of these other architects and engineers. In this process, we learned so much!

For example, we found that the fire-rated partition systems of that time were of gypsum block or of brick. As the building moved in the wind, the tops of the walls would move with respect to the steelwork above, creating a crack. Accordingly, the walls were excessively permeable; that is, driven by stack pressure, the walls allowed significant air flow in the building. At that time, energy consumption was not an

❺ Twin Towers, World Trade Center, New York

❻ World Trade Center. Spandrel wall panels

❼ World Trade Center. Tube structure

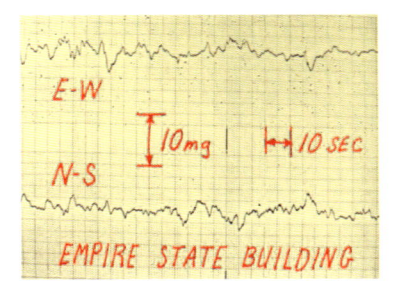

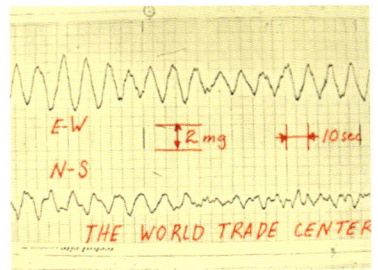

8 Empire State Building, New York. Oscillation trace

9 World Trade Center. Oscillation trace

issue, but, of course, those buildings consumed excess energy in no small part because of that air flow. Also, in the case of a fire, this air flow carried smoke to all of the floors above the fire. In response, we devised a new kind of fire-rated partition system, called Shaftwall, which changed the very nature of structures for high-rise buildings.

Let me explain. Driven by the wind, a tall building drifts more-or-less downwind, oscillating about its two axes. At that time, architects and engineers were not concerned with the level of oscillation of their buildings because the block or the brick partitions of that time added significantly to both the lateral stiffness and the structural damping. This trace (fig. 8), taken from the Empire State Building, depicts a highly damped structural system oscillating in the wind. Calculations compared to field measurements show that one part of the stiffness of the Empire State Building comes from the steel frame and about five parts from the effects of the in-filled masonry. Engineers did not have the tools for determining the effects of that masonry; instead, they learned from experience that the system performed well. With the introduction of Shaftwall, the structural systems of the past were no longer acceptable. Compare the trace of fig. 9, taken from the World Trade Center, with that of fig. 8, taken from the Empire State Building.

How does one deal with the absence of masonry? We knew that our building would oscillate in the wind but we had no idea of the magnitude of that oscillation. Fortuitously, two brilliant Danish engineers, Jensen and Frank, had come up with the idea of the boundary-layer wind tunnel. For the first time, in their wind tunnel, it became possible to replicate the real pressures and suctions on small buildings. Borrowing their concept, and, with the enormous input from Dr. Alan G. Davenport, who worked in our company for more than two years, we expanded that technology for use in the design of the World Trade Center.

We had a tower constructed atop the Telephone Company Building, immediately adjacent to the site. From there and from two other locations in lower Manhattan, we measured the turbulence of the wind at full force, comparing this turbulence with that which we found in the wind tunnel. That comparison gave us confidence that we could reliably use the boundary layer wind tunnel as a design tool. All of the above was coupled with a meteorological study to determine the characteristics of the gradient wind.

We then knew how much our building would oscillate, but we didn't know how much oscillation was acceptable. We turned to psychologists, who advised us not to worry, that people would accommodate. But, still, constructing a $1 billion project at that time (about 1970), we could not accept such views without a certain degree of skepticism. A variety of simulators were available and, while we rode several of them, none was able to replicate the frequency range of our building. In response, we developed two motion simulators, enabling us to evaluate how much the building should move; the first is depicted in figure 10.

While these various studies provided us with essential information, the need to limit oscillation to acceptable levels now fell squarely in the field of structural engineering. We developed and patented a viscoelastic damping system that successfully limited structural oscillation to design values (fig. 11); we were fortunate in convincing 3M Company to fabricate the dampers. The damping units, forming a structural system perpendicular to the primary system, consume a portion of the wind-induced energy of oscillation of the building.

The design of the aluminum and glass facade remained a knotty problem. We developed a new theory to establish the relationship between the strength of the glass, which is related to the rate of loading by the turbulent wind, and the statistical strength of the glass. In this way, we were able to develop breakage rates for various kinds of glass, arriving at the selection of a semitempered glass.

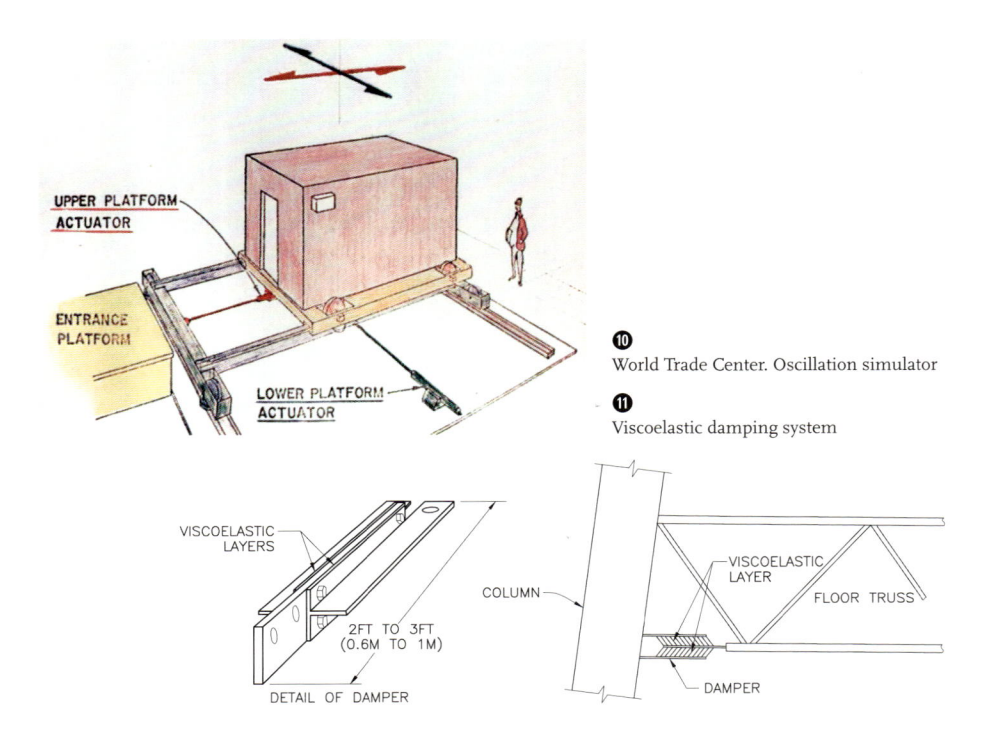

10
World Trade Center. Oscillation simulator

11
Viscoelastic damping system

But what if all of our structural designs had left undiscovered an issue of importance? Recognizing the lack of historical precedence, we looked for possible remedial measures, which would not be undertaken unless the construction and/or subsequent use of the building unveiled an unanticipated problem. For this eventuality, we developed the concept of outrigger trusses, a system employed commonly in today's buildings. These outrigger trusses, to be installed immediately below the roof, could provide additional stiffening and strength. And then came the need to provide for TV towers above the roof. Steel fabrication and steel erection had moved forward. We were able to realize the design of the outrigger trusses, which provided the anticipated stiffening and strengthening while safely supporting the TV towers. Fortunately, no other problem came to the fore.

Prefabrication was attained to an unprecedented scale. For example, individual pieces of steel plate for these wall panels (fig. 12) were fabricated in Japan, then shipped to Seattle for assembly into panels. They were then sent to New Jersey before being erected in the building. The entire process was nearly seamless. Why? We had taken all of the parts for the World Trade Center, almost every piece of the steelwork, describing them in IBM punch cards. For these panels, the size and grade of each of the plates, every weld, every bolt, all dimensions, were provided in punch cards. The basis of the contract with the almost forty steel contractors was not the drawings but the punch cards. Probably for the first time, the structural designs for a high-rise building were provided to contractors in a digital format.

The buildings, when finished, were opaque from most vistas and were appreciated more by artists and sculptors than by architects (fig. 13). When the sun is behind the buildings instead of in front, the airy lightness can be appreciated (fig.14).

12
World Trade Center. Prefabricated wall panels

13
Philippe Petit walking a tightrope between the Twin Towers, August 7, 1974

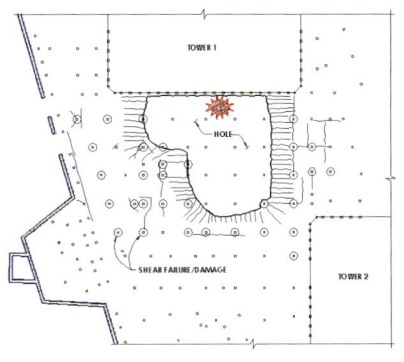

14
Twin Towers, with the sun behind them seen from New Jersey right after completion

15
World Trade Center. Plan showing bomb damage areas, 1993

Fortunately, early in the design, we had convinced Aaron Schreier (the project architect) and Minoru Yamasaki that it was unwise to allow parking under the towers. In February of 1993 came the bomb, which was set off about 2 meters outside of the south face of the North Tower. The perpetrators had the clear goal of bringing down the building, but the structural steel of the exterior wall was strong, tough, and stiff, reflecting the energy of the blast to the south. The bomb blew a hole in the reinforced concrete work outside of the tower (fig. 15), creating damage for a distance of about 100 meters. Little structural damage was done to the towers themselves. However, the towers acted as chimneys, sucking all of the smoke and airborne debris from the blast site up into the buildings, creating nonstructural damage and incalculable levels of human suffering.

The events of September 11, 2001, were devastating. Knowing that the Empire State Building had been struck by a Mitchell bomber, in 1945, we had designed the project for possible impact by a 707 aircraft; the design was for a low-flying, slow-flying 707. The 767s that struck the buildings were a bit heavier, but were flying at maximum speed, imparting significantly higher levels of energy into the building than had been anticipated. While the structures stood, one should not feel complacent. The Mitchell bomber, the Boeing 707, and the Boeing 767s are small when compared to the present-day Boeing 747 and to the new Airbus. Accordingly, I believe strongly that we should not design our buildings to withstand the attack of these heavy airplanes; instead, there are more fruitful ways to save lives.

This is not to say that the structural design has not received adverse comments, sometimes unfairly so. One noted structural engineer took it upon himself to criticize the structural design publicly because of a perceived lack of adequate anchorage of the columns to the floor trusses. Quoting from the TV program:

> They had two ⅝" bolts at one end of the truss, and two ¾" at the other end, which is perfectly fine to take vertical load, and perfectly fine to take shear loads, but once the floor elements start to sag during a fire, ... okay, they start exerting tension forces because it becomes catenary like a clothesline, and those two little bolts just couldn't handle it.
>
> As you start to lose the lateral support due to the floors, the exterior just crumples like a piece of paper or like if you took a sheet of cardboard, put some weight on it, and take out the lateral supports, it'll just bow right out.
>
> Had the floor system been more robust, with much stronger connections between the exterior and the inside, I think the buildings probably would have lasted longer. Would they ultimately have collapsed? Maybe not.

While that engineer had full access both to the drawings and to the site, he apparently neglected to look at either. In fact, the "two little bolts" were erection bolts, with the final connection by complete penetration weld. For Tower A, we designed for loads of twenty-seven times that of the normal practice (two percent of the axial force in the column) and thirteen times for Tower B.

US Steel Building, Pittsburgh

While designing the World Trade Center, working with the architects Harrison, Abramovitz and Abbe, our firm was awarded the commission for the US Steel Building in Pittsburgh (1970; fig. 16), until very recently the largest privately owned building ever constructed. We were responsible for the design of the tower, and Edwards and Hjorth created the areas outside of the tower.

The building, triangular in plan, is supported on perimeter columns located outside of the building envelope. The columns, being unclad and the structure visually exposed, are subject to changes in temperature from solar radiation, the atmosphere, and fire. The architects had proposed columns at a spacing of 4 meters, with the columns attached back to the building at every floor. Taking the contrary position, we suggested that two out of three columns be deleted and that the columns be connected to the building at every third floor. The proposal resulted in significant cost savings and drove the aesthetic

16
US Steel Building, Pittsburgh. Aerial view

17
US Steel Building. Pendulum diagram

18
US Steel Building. Corner column

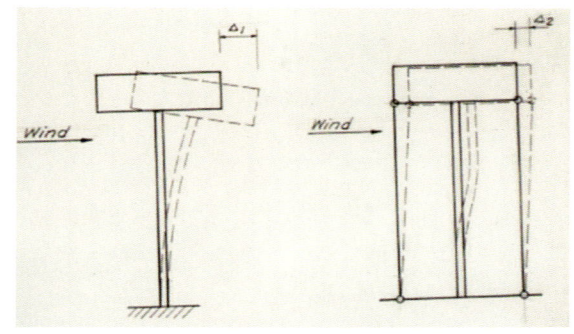

design of the building. Our logic stemmed from the desire for simplicity in the structural system, and from the fact that the connection of the columns to the building required the penetration of the facade wall (creating issues of structural vs. architectural tolerances, watertightness, and the like).

The concept of the structure is easy to comprehend (fig. 17). Imagine a rocking pendulum with a mass at the top, visible on the left in this image. We borrowed the idea of the outrigger trusses used for the World Trade Center, tying the tips of the outrigger trusses to the ground. These trusses dealt effectively with the thermal expansion and contraction of the perimeter columns. While they were a simple idea, the outrigger truss worked extremely well. The structural system, then, is that of a braced core building with the exterior columns acting as tension and compression members, a concept that is used to this day.

This is one of the corner columns in the services core (fig. 18). The center plates are 200 millimeters thick, of Grade 50 steel (345MPa), while the perimeter plates are 100 millimeters thick, of Grade 42 steel (290MPa). This hybrid design of the column is almost as strong as if all of the steel were of Grade 50 material.

19
US Steel Building. Steel strength diagram

20
US Steel Building. Cross section

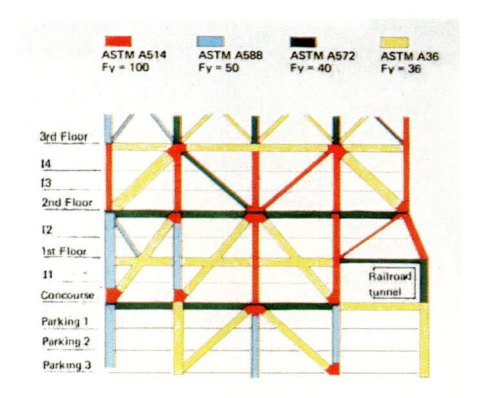

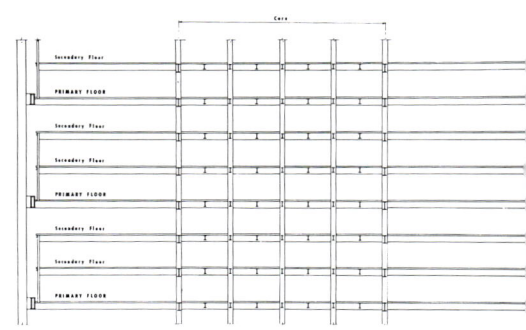

It is customary to think of the use of high-strength steels as carrying higher loads but, in many instances, we used high-strength steels to avoid carrying loads. That is, the use of high-strength allows a member to accept a large deformation, i.e., a larger stress (fig. 19). In that way, we move a given load to a more desirable location. The cross section of the building (fig. 20) shows that the perimeter columns are attached at every third floor and that two intermediate floors are posted down to the primary floor with 15-millimeter-wide flange posts.

Because there was no cladding or conventional fire protection, the columns were liquid-filled to provide fire safety. Due to corrosion inhibitors and antifreeze, the liquid has a specific gravity of about 1.2. Divided vertically into zones, the columns of any one zone are fully interconnected. Within each zone, the liquid circulates freely, up in some columns and down in others, all driven by differences in the temperature of the liquid. It being a completely passive system, there are no pumps.

The roof was designed for vertical takeoff jet aircraft, but, as far as we know, no such aircraft has ever been off of this roof. We studied various techniques to control the turbulence of the airflow over the surface of the landing field. For this purpose, the PanAm Building in New York used turning vanes at the parapet; PanAm had an operating commercial heliport for many years. We were able to demonstrate that these turning vanes were not effective and that a normal parapet detail was quite satisfactory.

21
AT&T Headquarters, New York. Structural diagram

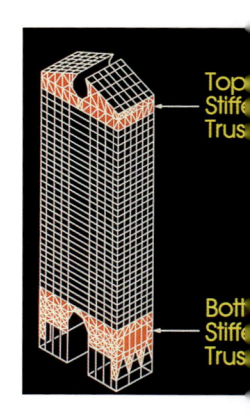

AT&T World Headquarters, New York

With the architects Philip Johnson and John Burgee, we obtained the commission for the AT&T Headquarters in New York City (now the Sony Building). In that it makes use of every odd corner to contain the lateral-force system, the structure is typical of New York. The perimeter consists of two C-shaped rigid frames around the short sides of the building connected to a soft-rigid frame down the center of the long sides (1958; fig. 21). Bracing structures are carried across the top and bottom so that the two C-shaped rigid frames act as a single structure. Since this is a very slender building, we added interior outrigger frames. At the base, steel-plate shear walls surround the services core.

Bank of China Tower, Hong Kong

Another breakthrough in my life and that of our company was the Bank of China Tower in Hong Kong S.A.R. (1990; fig. 22). I. M. Pei, the designer of the project, fully realized that the site was next to the Hong Kong and Shanghai Bank designed by the very talented Norman Foster. This wonderful, very elegant building cost about three times more than was allowed by our budget and for the same floor area. Certainly, Pei was not prepared to develop a design that was inferior to its neighbor. Accordingly, carrying the project in his mind for about a year, he developed this fascinating design. Only then did he accept the design commission. Pei gives me far more credit for the design than I deserve.

The original concept (fig. 23) placed Vierendeel-like trusses at what are called refuge floors, where people go in case of a fire; the trusses were visually accented in the facade. The executives of the bank drew a line in the sand, stating that the structure was wonderful but that it could not be expressed in the facade. They would not tolerate Xs, the symbol of death, on their facade. With his great talent, Pei came up

㉒ Bank of China Tower, Hong Kong. Exterior

㉓ Bank of China Tower. Original concept

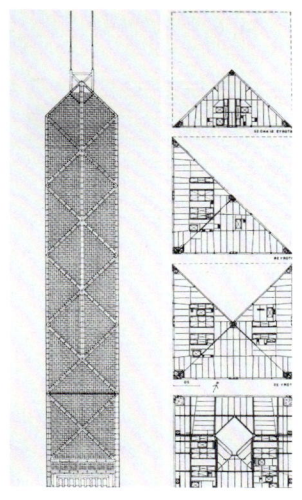

with this design (fig. 24), and with his golden tongue he was able to convince the executives that the resulting diamond was deserving of being expressed in the facade.

The building's 54 meters at the base drops off a quadrant at a time as it rises to a height of 369 meters (fig. 25). Under the action of lateral forces, the geometry creates a large torsional eccentricity. By judiciously designing the bracing systems, it was possible to minimize both the steady state and the dynamic consequences of that eccentricity.

Those of you who have worked with Pei know that everything has a reason, a starting point and a stopping point. The architectural representation of the diagonals stops at the very corners of the facade (fig. 26); of course, the columns must be inside and must consume an appropriate amount of space. Our approach to this dilemma was to use plain frames in structural steel with two different geometries, it being a space-frame structure. Then in the corners we knitted the columns of the space frame together with reinforced concrete (figs. 27, 28).

The problem with this approach is the transfer of loads from one plane frame to another and to the enfolding concrete; clearly, the various vertical elements are eccentric from one another. In essence it becomes a study in eccentricities. Still, we were able to demonstrate that we could build with all of these eccentricities without creating bending moments in either the steelwork or the concrete work. The structure was so designed, and stands stalwartly today.

Hong Kong, like other places prone to hurricanes or typhoons, is subject to high-wind loads; "typhoon" and "hurricane" are differing names for the same weather phenomenon. For a hypothetical 304-meters-high building, this graph depicts the wind and earthquake loads (in this case, base shear) for various places around the world (fig. 29). Perhaps the most interesting lesson to be learned is that the

㉔ Bank of China Tower. Final design

㉕ Bank of China Tower. Floor plans

㉖ Bank of China Tower. Facade, close-up

㉗ Bank of China Tower. Corner columns knitted together

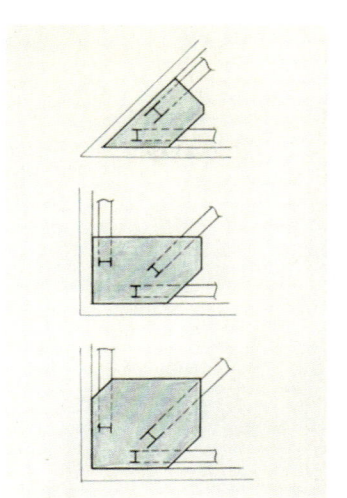

㉘ Bank of China Tower. Corner columns in place

㉙ Base shear graph for various locations worldwide

earthquakes of Los Angeles, or for any other location on the earth, produce rather lower lateral loads than are produced by the wind in Hong Kong. For a very tall building, the wind loads exceed the floor loads. In short, for a very tall building, the wind loads tend to drive the conception of the structural system.

This necessitates a return to a discussion of the wind tunnel. There are three different kinds of wind-tunnel tests. The first is a pressure model, designed primarily to measure the steady state and the fluctuating pressures on the facade (fig. 30). This same model is commonly used to evaluate the street-level wind environment. The data is of use to the architect, the specialist facade designer, the building services engineer, and the structural engineer. The second is the force-balance model (fig. 31), which, like the pressure model, does not sway or oscillate. Instead, it is attached to a dynamic force balance that measures both the steady state and the dynamic components of the overturning moments, shears, and torsions. These data are of interest to the structural engineer. The third is the aeroelastic model, which is far more complex than the others. It is a five-mode model (fig. 32) that is

㉚
Bank of China
Tower. Wind-tunnel
pressure model

㉛
Bank of China
Tower. Wind-tunnel
force-balance model

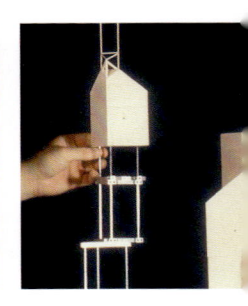

㉜
Bank of China
Tower. Wind-tunnel
aeroelastic model

(above) ㉝
Bank of China Tower.
Barn structure at top

㉞ and ㉟
Bank of China Tower
under construction
and lasso

80 | THE FELIX CANDELA LECTURES

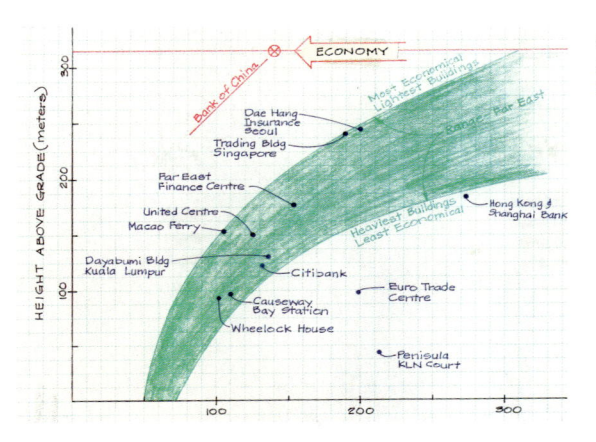

36
Bank of China Tower. Economy graph

designed to oscillate with the same characteristic frequencies and mode shapes that exist in the real building.

Pei came to us, unhappy with the ambience of the space at the top of the building, saying that he wanted a "barnlike" structure. Accordingly, we scratched our heads, turning to the works of Bruce Goff to sort out the nature of barnlike structures. However, the translation from Goff to structures of this scale did not work well. In any event, we came up with this idea (fig. 33)—erecting two masts at the top of the building by telescoping the sections and then by jacking.

As you can well understand, the bank was very concerned about the cost of the steelwork. Most politely they told us that the Macau Ferry Building had less steel per square meter than ours did. Nippon Steel Company assisted us by putting together this chart (fig. 36) and plotting the unit weight of the steelwork against the height of the building. The chart demonstrates that the Bank of China was very economical compared to other buildings in the Far East. The structural steel for the building was fabricated in Japan and shipped to Hong Kong. The steelworkers often handled the steel with remarkable bravery (figs. 34, 35).

In my view, this is a most beautiful building. It has aged extremely well, looking as good today as it did when it was first built.

Meyerson Symphony Hall

We worked with the same architect, I. M. Pei, on the Meyerson Symphony Hall, in Dallas, a project by the architects Pei Cobb Freed & Partners (1989; fig. 37). The long glass wall became the topic of discussion because the longest of these mullions was about 21 meters in length. I explained to Pei that the uneven deflection of the mullions, particularly under wind load, would be disturbing and that a bridging system would be required to hold it together. While not wanting to prejudice him, I drew a sloping line across the drawing of the facade to depict the bridging. The brilliant Pei, ever able to change a liability into an asset, came up with this absolutely gorgeous design.

Here, the bridging, with the sloping lines, is seen from the inside (fig. 38) as it wraps around the facade. The bridging and the structural mullions are covered with a sunscreen through which you see the structure. It is essential that the structure be clean and elegant in order to be accepted by a master such as Pei.

37 Myerson Symphony Hall, Dallas. Exterior

38 Myerson Symphony Hall. Interior

Sloping Buildings

Working with the architect Gunnar Birkerts, we came up with a structural design for an unbuilt sloping building in Michigan for the Domino Pizza Company. In jest, we called it the Leaning Tower of You-Know-What. Indeed, it was prodding on the part of Birkerts that led us to think through many of the problems of the sloping building. These problems include the realms of materials technology, structural analy-

39 Madrid. Sloping buildings

40 Madrid. Sloping buildings under construction

82 | THE FELIX CANDELA LECTURES

sis, and design and construction methodologies. Probably more than I am, Birkerts is a careful thinker, not one to rush into new arenas without cognitive thought behind the design. In any event, he led us through the process while, at the same time, creating the opportunities for our holistic approach to the structural design.

Later, with John Burgee Architects of New York (and Pedro Sentieri in Spain) we were retained to provide the structural engineering for two buildings in Madrid, one on each side of the street (figs. 39, 40). Except for concrete cores, these are all-steel buildings; the lean is about fifteen degrees. The structural systems, both those in steel and those in concrete, are post-tensioned to compensate for the cantilevering of the building.

Miho Museum Bridge, Shiga-Raki, Japan
The site selected for the Miho Museum was in a forested nature preserve located in the rugged mountains to the east of Kyoto, Japan. The architect I. M. Pei conceived of the access to the museum as a tunnel through a mountain, linked to a bridge spanning 200 meters over a

㊶ Miho Museum Bridge, Shiga Raki, Japan

㊷ Miho Museum Bridge. On deck

wooded valley (1997; figs. 41, 42). Most of the visitors to the museum travel by foot, placing an emphasis on the need for an aesthetic experience both in traveling over the bridge and in viewing it from the side.

The structural design of the Miho Museum Bridge is intended to integrate artistic beauty and structural elegance; post-tensioning finesse, small structural elements, and an innovative drainage system achieve the objective of a light and airy structure. This being a post-tensioned design in structural steel, the bridge was constructed below the final grade before being raised into its permanent position by the arch-supported post-tensioned cables radiating from the tunnel and by a king post located beneath and toward the museum end of the bridge. Making use of the tunnel as the root of the cantilever, the all-steel bridge is fixed against translation and rotation at the tunnel and hinged, but allowed to move longitudinally, at the museum. The combination of end fixity and post-tensioning produces high axial loads at the level of the deck, but low axial loads at the single bottom chord. The result is a reduction in the diameter of the largest pipe sections of the truss to only 25.4 centimeters. Making use of existing technology for tennis courts, the walking/driving surface of the bridge is composed of a porous ceramic infill, supported on stainless steel grating. The system allows the elimination of bridge drains, which would have been larger than the structural members.

World Financial Center, Shanghai
This is a project in Shanghai with Kohn Pederson Fox as the design architects (fig. 43). The Mori Building Company of Japan is the developer.

This building, having passed through the hands of several structural engineers prior to our involvement, was to be 450 meters high. When the design came to us, the foundation piling had already been installed. Our charge was to increase the height of the building from 450 to 492 meters and the floor area by fifteen percent, while making use of the existing foundations. This could be done only by creating a lighter and more efficient structural system.

We developed this design, which is reminiscent of a pair of bent leaves (fig. 44). In need of additional stiffness, the leaves are braced across the top, much as was done for the AT&T Building. The original

43
World Financial Center, Shanghai. Rendering

44
World Financial Center. Design comparison showing earlier design with circular opening at top

㊺ World Financial Center. Under construction

design included a moment-resistant space frame with closely spaced columns of considerable size. Seeking to improve the views from inside the building, we chose to use not more than three columns on each face, keeping the columns small by making them load-bearing for a maximum of twelve floors.

We made use of three-story-high outrigger trusses. Unlike conventional outrigger trusses that pass through the services core, we chose to pass them around it, burying them in the perimeter walls of the concrete services. Outrigger trusses are complicated, difficult to fabricate, and time-consuming to erect, and hence expensive. Even so, their use contributed significantly to the stiffness and strength of the structural system. We believe that this building achieves the highest level in economy, strength, stiffness, and robustness. Perhaps of more importance, we believe the architectural design is most elegant.

The body of work discussed herein is only a small fraction of the wonderful structural designs developed by the men and the women of Leslie E. Robertson Associates, R.L.L.P. No one person should be singled out as the creator of the structural design of any of these magnificent works. Indeed, without the strong input, steady hand, and creative genius of the architects, these buildings and bridges would never have been realized. We remain ever grateful for the opportunities to work with these creative designers and with the high-tech architects who work with them.

❶ Felix Candela. Church of San José Obrero, Monterrey, Mexico

Shell Structures: Candela in America and What We Did in Europe

HEINZ ISLER

Felix Candela was, for his time, the master builder of shell structures. No one else realized as much as he did in this area. He brought a kind of mission to the world of building. His structures have a lightness and elegance that had never before been achieved. I think everyone who knows his work, which is so thin, light, resistant, and beautiful, can agree.

What is amazing, and yet not amazing, is that he was architect, engineer, and contractor all at the same time. This was essential for what he achieved. Anyone who works as a builder must have within himself a sense for all aspects of a building, from the architectural design to the engineering and statics, and even the construction and beyond. The construction is just one aspect. When it is finished, the life of the building begins. This life should be not just a few years but three generations, about three times twenty, or sixty years, without any damage or need for repairs. That is the standard we should have, but in practice it is difficult to name ten structures in which this standard has been achieved.

We, as young people at the time, admired Candela's work and were encouraged to think along similar lines. This is the influence he had on new generations. The building that had the most influence on my development was the restaurant in Xochimilco (1958; see page 168), which is a shell of repeated regular hyperbolic paraboloids with an extreme thinness of only 4 centimeters. The shell has a tremendous span and the interior of the pavilion is a work of art. Two other wonderful Candela works are a church with two wings, the Church of San José Obrero, in Monterrey (1959; fig. 1), and an open chapel in Cuernavaca (1958; see page 168). Both of these are masterpieces, especially the latter because Candela had very interesting experiences while building it. The famous open-air dance hall in Acapulco (1957) has, sadly, been torn down in recent years. I think society should unite to save these masterpieces from destruction. With money and other material things to motivate them, people can lose respect for built works.

Moving away from Candela's own work, I wish to discuss his influence in the United States and what was being built in Europe at the same time. Many shell structures were created in America. One example is the famous TWA Terminal at John F. Kennedy airport, built by Eero Saarinen in 1962. It is in danger of being torn down for practical and organizational reasons, but there is opposition today, and people are uniting to save it from destruction. It is a beautiful object that is more like a monument but also offers practical functioning space for its intended purpose. The St. Louis Abbey (1962), for which Pier Luigi Nervi was a consultant, has three layers of repeated shells. The famous airport in St. Louis is another great example of a thin shell structure (1956). Anton Tedesko, the engineer, collaborated with the architect Minoru Yamasaki and made the necessary calculations for the shape to be built.

Of the many techniques for constructing shells, a number were invented in the United States. One such method, which was later used frequently, is a balloon that is used as structuring. A hardening material is poured on the doubly curved balloon and when the material has fully hardened, the balloon can be removed and a shell is left. This method is very easy, but not the most economical because of the high cost of the balloon. My office has worked with these balloons, and has found that the same balloon must be used at least twenty times to realize a considerable price reduction. Another method of shell construction can be seen as an ingenious device for creating a reservoir. In this example, the client wanted a dome without the expensive formwork, so a hill of earth was created, forming an adjacent reservoir. The form was solidified on top with some water, and then the concrete was poured on. When hardened, the shell was lifted into place and the earth was removed by truck.

In Europe there were similar developments. A work by Eduardo Torroja, the Algeciras Market, built in 1934, already shows the characteristics of a modern, beautiful shell. It is more than sixty years old and remains undamaged. I think this is an example of how things should be done and could be done. Torroja's Zarzuela Hippodrome in Madrid (1935), with its cantilevering shells, is also a very audacious design.

An example of statics from France is the Royan Market, created by architects André Morisseau and Louis Simon with the engineer René Sarger. Both the exterior and interior are good examples of what was possible at the time when it was completed, in 1956. In Germany there was a great deal of construction, one example of which is a dome for an aircraft hangar. Prefabricated pieces were joined together to form the dome shape. In the eastern part of Germany, Ulrich Müther was the engineer and contractor of the famous Teepott restaurant in Rostock, completed in 1968, with architect Erich Kaufmann. One again sees the unity of these two professions, engineering and contracting. Another very interesting approach is a shell shelter made solely of timber. Straight slats are nailed together to form the shell shape. When there are two together, the shape is still mobile and flexible, but as soon as a third strip of wood is added, it becomes rigid immediately and can serve as a good shell.

I consider the Italian architect Pier Luigi Nervi to be the master of shell structures. He built numerous very beautifully ribbed domes, using prefabricated rib elements to construct economically competitive shells. He also utilized cylindrical shells, one after the other, that have considerable spans of about 20 meters. By avoiding a heavy edge beam, he tried to show the lightness of the construction.

When I was a young engineer, I was very fond of these light structures. At first I was against building because I had seen how it could destroy nature. I was of two minds: on the one hand, I didn't want to be part of this destruction; on the other hand, I was fascinated by doing things very elegantly and very lightly. I wanted to investigate thin shell structures. When I became a practicing engineer, one of my first buildings was a factory, a shell structure with about a 22-meter span. Without knowing it, I used two new techniques that did not exist before. The first was the free edge, a straight line of 20 meters' span with a thickness of only 12 centimeters. The second was a prestressed shell with two cables installed within it, one embedded at the top of the shell and one at the bottom. Thus I completed my first shell based on purely technical and not aesthetic or practical considerations.

Then came my suffering phase. I was assigned to work on a concert hall in a village near Bern. The architect wanted a barrel shell on top with rounded ends and straight sides. I was uncomfortable with this request. The straight line is, for me, a curve with an infinite radius, and I knew from my experience that shells live or die with curvature. When you have no curvature, your ability to carry loads decreases to two percent—a fiftieth of the total. When you have good curvature, you have a capacity of about seventy percent, and up to about ninety percent if it is very good. This is a tremendous difference, so I decided to make a curve at the edge instead of a straight line. I began to design on my drawing board—I think at the time not even the word "computer" existed—doing everything by hand with small curved instruments. I tried to find a shape on a rectangular plan that had an overall curvature—not straight in one place, with a radius or circle elsewhere. It was easy to do it with one section, and another section, and maybe a third section in a diagonal. But when I began to elaborate for the formwork, I had to give the coordinates for all these curves, and no two curves were alike. As soon as I began, it seemed to be an insurmountable task. I had 400 points to define the shape that I wanted, and when I had to change one point, all 399 other points changed as well.

As you can imagine, I might still be working today on this absolutely impossible task, made worse by the fact that the building was progressing as I tried to keep up with it. I would work till two or three o'clock in the morning so that the next day I could deliver three new curves to the builder. I worked day by day, week by week, and finally the job was completed. My biggest concern was that the last girder would not go together with the first, creating a step in the shell. But that did not happen, and after more than forty years, from a technical point of view, the shell is still stable. For the past fifteen years, fifty tons of gypsum have been hung from the ceiling in order to make an internal landscape for dancing, and the structure is still sound.

Personally though, after this building was finished, I was disappointed. I was not at all proud, but only dissatisfied. I had not achieved my aim. I had attempted an impossible task that could not be realized. I thought surely that was my first free-shape structure and also my last.

But then I made an observation. Something came to me, at three o'clock in the morning, when I was standing in my bedroom. I just wanted to fall into my bed, terribly exhausted and weary, when something caught my eye. A few feet in front of me was the solution to my problem. The pillow on my bed had the exact shape that I had been looking for for weeks, months. In my mind the shape began to rotate, and I saw that it was the solution to the problem that I, as an engineer with six years of training and diplomas, had been unable to find. It was clear to me because I was trained in doing model tests for any problem that could not be solved by doing calculations. I saw the way to build a pillow, a technical pillow of the highest possible precision. Of course, precision is an absolute necessity when you build something with your hands, with models.

It was too late for the concert hall but we were working on a new project that would benefit from this discovery. Within a few months we had made a pillow to test my idea (fig. 2). It was a rubber membrane,

❷
Wood frame holding a rubber membrane, inflated to make a pillowlike form

Interior of the pneumatic-form roof of the Eschmann Chemical Factory, Thun, Switzerland

continuous and extremely precise. This balloon was in the lower frame, clamped together so that the air could not escape. Another frame was attached on top, constructed of very old, hard wood that would no longer deform. A measuring device was attached to the top wood frame that allowed the balloon to be measured. The pneumatic apparatus was merely a bicycle pump, neither difficult to use nor expensive.

There were only two small problems. The first was that I could not use a needle to measure the shape of the balloon precisely because the needle would puncture the balloon. I could only test it by using a needle so thin that it was weightless. The second problem was that the balloon would leak. You cannot build anything in the world that will not leak. I had anticipated this leaking, but I was confused when, instead of losing air, the membrane became larger as I tried to measure its shape. After some time I realized that I was acting as a heater when sitting near the apparatus, warming the air and causing it to expand. The solution I found was to wear my winter coat and gloves with the fingertips free, and cover my mouth so that my breathing did not affect the model. The balloon stayed still and I could measure it to a fraction of a millimeter at every point. This is a good example of what can happen when you begin to work with physical models. Ultimately I could measure all of the curves from the membrane and build the wooden scaffolding for the shell based on these curves. The first hall built with this method covered an area of 23 by 14 meters, only needing pillars at the edges of the shell shape.

On top of each shell is a hole 5 meters in diameter to get a very good light spread into the building. It is covered with a dome of milky glass reinforced with polyester, so that the light is diffused, which is very good indoors. From then on we made many similar shell structures for factories and garages.

These shells could be further refined. For my early shell projects I had worked with architects, who were just the architects of the village or the town, and I was always rather dissatisfied because they didn't understand what I wanted or what was possible. Sometimes they installed windows that were very ugly or pieces that were too heavy, affecting the lightness of the structure. I decided that when I next did a shell, it would be on condition that I control everything. Having made this decision, I began to do projects where I worked more with the structure itself.

The largest shell that we built was 60 by 60 meters, for a dispatch center in a big factory. For interior lighting in this shell there are seventeen holes, from 4 meters to 4.5 meters in diameter. You can easily put holes in a good shell because the stresses flow around them without any interference. A shell such as this works as if there were four arches that are inclined toward each other.

We built hundreds of shell buildings over the years, one after the other. In some we used blocks of shells together to form big halls, one with twenty-nine shells that had a maximum inside volume with minimum cost. More recently my office has built shells of about a 40-meter span, that is, there are 40-meter spans from pillar to pillar, some 1,600 square meters of self-supporting roof without any ribs or inside reinforcements. It is just a membrane, but with a specific membrane shape the job can get done.

About a year after my observation with the pillow, I came across something at a building site. I was standing on top of a shell, walking across the reinforcement, and suddenly something struck my eye. As seen in this original photo from 1955, I saw burlap hanging in the

❹
Wet burlap cloth hanging in tension over wire mesh

5
Model of gardener's netting with ice in Isler's yard, Zuzwil

meshes of the reinforcement (fig. 4). This burlap was wet because it had been raining the night before, and the sun was reflecting on the hanging shapes. What I saw was amazing to me. It was again the shape I had been looking for: a square frame, with a membrane hanging from it in a natural shape. This was even more basic than the pillow and the blown-up membrane; it was just hanging by itself with no apparatus. I was very enthusiastic about this new discovery, and I began to see the possibilities in my mind. I quickly took a photograph of my discovery and rushed home to experiment.

This is a shape that has a contour frame but can also be hung on points. I constructed a model consisting of gardener's netting of 4 by 4 meters hanging on four posts, creating a point-supported shape (fig. 5). It must be precise and be supported very well, and within half a second, the shape is created automatically. Nothing has to be done; it just has to hang. I wanted to harden the netting and used a material that is not very expensive: ice. I soaked the netting in water and on a winter day it was frozen in a quarter of an hour. I couldn't believe it was less than one millimeter thick, with a span of 4 by 4 meters. It was so stiff that we could turn it over. After fixing the distance of the post supports with some slats, we carefully turned the shell upside-down and put it down onto the supports. I knew it would work, but I did not know it would work with such a small thickness. This experiment, along with others, was continued, and much more was learned with time.

We developed a standard technique for making these kinds of models for various projects. An example is a model for the covered open-air theater in Grötzingen, Germany, with a front span of 42 meters, on a slope with five supports (fig. 6). We have a device for measuring the

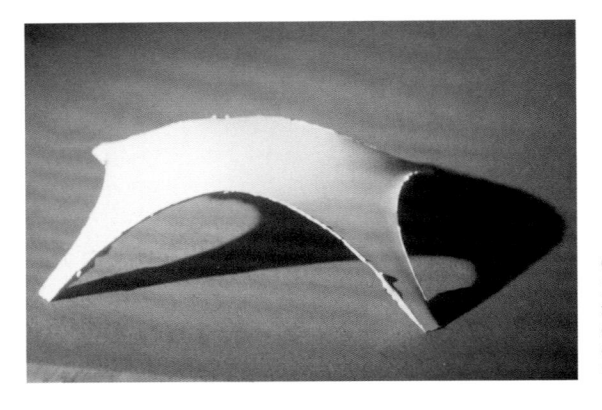

6
Reversed-membrane model, supported on five sides, for the open-air theater, Grötzingen, Germany

models in order to transfer the points of the curves to the construction site. The model of the structure is placed in the device, and a rod measures the curves with a measurement reading that has an accuracy of about one-fiftieth of a millimeter. This precision is necessary in order to build the scaffolding correctly.

Working from models, the architect Michael Balz designed the open-air theater at Grötzingen, which was finished in 1977 (fig. 7). It sits in the slope with a lightness created by the slender edges that are upturned slightly, therefore avoiding the use of an edge beam. This is a continuous surface that cannot be defined mathematically or by anything else; it can only be shown in rows of numbers. Today the computer can help smooth the curves and take sections of the shape and so on, but the principle is that the model by itself creates a natural shape through gravity.

From the first experiment with the hanging cloth and ice to the precision of a model used for construction was a long and difficult course. There were questions about which materials to use and how to handle them, and there were other problems as well. But finally I developed a process to make models that are precise at every point to at least a twentieth of a millimeter. They are so precise that I can take a model and translate it point by point to the actual size. This took patience and time, but it can be done.

I think that these projects represented more than just completed works. I have mentioned two of the methods that I used: the inflated membrane and the hanging membrane. In principle, I discovered methods that did not exist before those years, 1954 and 1955. Antonio Gaudí had created curves that define his wonderful churches, but these were curved linear elements. I do not know if the surface of a hanging shape had been investigated before, in order to make real, precise, and accurate buildings. Through these two observations I discovered a new category of shell construction, which I like to call the structural shells. They follow a natural law, a structural logic. The forces in the shell travel in the most direct way that they can to reach the ground. Whenever you construct something, using the most direct approach to it will be the cheapest, safest, and statically the best way to build.

These methods allow point supports to carry the entire load, sometimes up to an area of 2,000 square meters with only four supports. It is a continuous surface, the surface itself. There is no need for cladding or finishing of the surface, and it is not just a supporting structure but is also space-embracing. Instead of having numerous elements fixed together, one element performs all of the essential functions. By choosing the shape correctly, so that it is only in tension, you will have only compression when the shape is inverted. There are a large number of these shapes that produce only compression forces. When you use a material, like concrete, that is very strong in compression and very weak in tension, choosing a shape merely in compression solves another half-dozen problems. The concrete itself is artificial rock. It is as hard and as durable as rocks in the high mountains and can last a hundred years, maybe a thousand years. When this artificial rock has a shape that places it only in compression, it is durable. There are no cracks, there is no rust, and there is no deterioration, so it needs no repairs for a long time. I think there is nothing cheaper than such a roof. This will not be true on the first day; initially it may be twenty or thirty percent more expensive than any other form you choose. But when you can wait ten, twenty, thirty, or more years without any maintenance, it becomes very economical.

I think if economy has its place, then ecology has one as well. I am careful not to abandon or waste materials, which is very good ecologically.

❼ Michael Balz. Open-air theater, Grötzingen, Germany. The inclined shell was designed by the hanging-membrane-reversed method. There are no edge-stiffening members.

❽ and ❾
Indoor tennis center, Heimberg, Switzerland. Under construction and completed

But the main point is that I think we are sick of the cubic architecture that prevails everywhere in the world. In Manhattan the towers have grandeur and character, but modern architecture that is just a box, and another box, and another box, is not pleasing. Some buildings have been painted to seem oblique, but that doesn't improve the primary quality of the architecture, which is volume. We need volume, and to play with the volume we need statics engineered to make it hold its shape. This line of thinking, the thinking that led me to the structural shells, creates an opportunity for designers, giving us more than just cubic buildings all the time.

Double curvature is the basis of every good modern structural shell. The load follows the shortest route to the foundation with very low stresses inside. An important fact is that a hanging membrane is in

96 | THE FELIX CANDELA LECTURES

10
Light scaffolding supporting wooden boards, Leuzlinger Sons Company

11
Laminated, reusable wooden arches give the form to the concrete shell of the Heimberg pool roof, Heimberg, Switzerland

tension, and tension has no instability. When this shape is turned upside-down, a new problem arises: a compression shape is formed, and a thin compression element will buckle. The models by themselves made that clear.

We built a factory and office building for the Sicli Company in Geneva, which was completed in 1969. It was a large-scale project of about 1,000 square meters with seven supports on irregular ground (fig. 15). Because it was only one application, we used flexible girders. The main work comes in defining every one of the curves initially. Every curve is different from the other, but it is just basic work for the laboratory and the mind.

Two important developments of these structural shells were the test model and the documentation of the shell's behavior over time (fig.

⑫
Swimming pool,
Brugg, Switzerland.

17). When this construction was underway, it was not possible to analyze the shells with a computer. The computer existed but it had no symmetries in the chosen shape, which would have surmounted all of its capacities at that time. So we created a way of testing the models systematically. A model of the shell had the assumed regular and pre-stressing loads applied to it and then the displacements were measured, point by point. Once the building was finished, and the raw shell stood, there were about 100 points that correlated to the model that we could measure.

An engineer can only measure one thing over a long period of time and that is the height. A length or a position can perhaps be measured today with satellites, but at the time only the height could be measured. The measurement can be taken in the first year, the tenth year, and the fiftieth year, remaining very accurate to a fraction of a millimeter. A length could not be accurate, because in fifty years the measuring rod will have changed length. But when we measure the height, we measure the deformations of the shell. Compiling the measurements from the first day, the first month, the first decade, and so on offers very interesting information.

Nothing was known about the behavior of this kind of structure when we started building these projects, but I had the concept, which I had learned at the university, of watching the lifetime of a building beginning the first day. It is going to deform and change; in ten years it is a different building from what was built originally. We measured structures that we built, and compiled data over thirty years. Taking a point at the first day, and then as the temperature changes, it goes up

13 and **14**
Wyss Garden Center, Solothurn, Switzerland

15
Factory and office building, Sicli Company, Geneva. The thin, continuous shell is on seven supports.

16
P. Wirz. Kilcher Company, Recherswil, Switzerland. The form was created by the flowing method.

ISLER | 99

17
Structural study and test model of the Sicli Company building

and down for the first week, and over the first year; in winter it goes down but in summer it goes up. It is always moving, every half an hour. You can smooth these curves and see the envelope over time. For instance, a point cannot go down and down and down; that would indicate a collapse. Another point is not allowed to go up and up and up, or it would break.

We found that these curves were all asymptotic, that they came to an end. Sometimes the final height difference was 30 percent of the deformation on the first day, and for others it was 100 percent. At a maximum, we had a structure with up to 300 percent of deformation compared with the first day of construction. These measurements are usually not taken at all. But when we can look at our building and know that it is coming to an asymptotic behavior, after about five years or twelve years or even fifteen years, only then can we be sure the building is safe and sane and can endure. We calculated these factors in advance, but we had no proof, so this monitoring is very important for new shapes. I have a shape and know how it is behaving over a period of time, and for new shapes it can be used as an example.

I want to repeat the methods of finding shapes. The first thing you can use is geometry—mathematics. Geometry means that the form has a geometrical aspect. For example, it might be a cylindrical shell, a rotational shell, or a hyperbolic shell. These shells are what I call the analytic shapes, and practically everything up to the 1960s was designed in this way. Then what came from our experiences were the experimental shapes: the hanging shapes; the pneumatic shapes; the flowing shapes that I have not touched on here; and the hanging reversed shapes. These are what I call the structural shells because they follow the forces that nature needs, that physics needs, that statics needs; and therefore they are far better than the others. The others are good and useful, but these can lead to excellent results.

What nature has done for us may be summarized as follows: for 200 million years it has created an immense production in a variety that is indescribable. In the sand, you find millions of tiny shells, down to a microscopic scale. Then there are thousand-meter-high rocks, built up as a bank of chalk. This chalk is just little fragments of small shells, hundreds of kilometers long and dozens of kilometers wide. It is an absolute mass production that has been on the earth for 200 million years, and I think we can say it is the most widespread construction that exists on our globe.

In a few weeks it will be spring. Open your eyes and look at what every leaf, every petal, really is. It is a doubly curved shell: it is round; it has curvature in every direction; and it is supported on one point. This is something that we engineers have not yet solved. These forms of nature are here for just a short time; they open, then shut. They are sometimes white like the cherry trees, or colored like the tulip, and after a week they fall to the ground. But they are not wasted, for they build new soil for the next spring to make new flowers. I would thus say that these forms are even richer in quantity and quality than hard shells. Nature is prevailing; it is an absolute infinitude on our earth. Just open your eyes, and when spring comes, appreciate it.

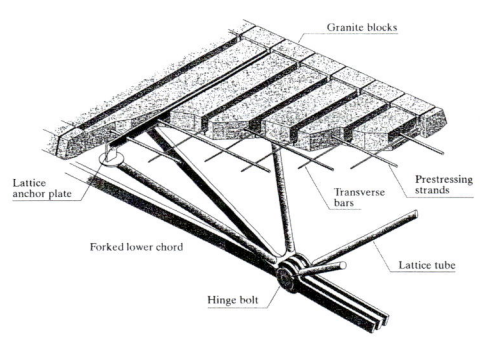

❶
Inachus Bridge, Beppu, Japan

❷
Schematic structural constitution of the Inachus Bridge

❸
Ceramic ornament at a joint of the Inachus Bridge

102 | THE FELIX CANDELA LECTURES

The Design of Structures— From Hard to Soft

MAMORU KAWAGUCHI

In 1992 Felix Candela and I, with some of our colleagues, were invited to Seville to give lectures at the Escuela Técnica Superior de Arquitectura de Sevilla in memory of Emilio Piñero, the great Spanish genius who developed many interesting spatial structures, including deployable ones that are popular today in space technology. In Felix's talk, he stated that structural design should be worked out and enhanced to such an extent that nothing could be added to it or subtracted from it. This statement made a strong impression on me, and, even today, whenever I design a structure, I ask myself if I am elaborating on my design to the extent he suggested.

I would like to discuss some of my works in memory of Felix. In doing so, I have rearranged them so that I begin with the hardest material, which I believe is stone, and end with extremely soft material, which is fabric without prestresses.

A Pedestrian Bridge of Natural Stone
The first example is a small footbridge that I designed in 1994 for the city of Beppu, on the island of Kyushu. When the mayor of Beppu asked me to design this bridge, at a very beautiful location, he laid down three requirements. The first was that the design should be unique. He said he did not want to find the same design elsewhere in the world. The second was that it should include some kind of suspension principle, because he liked suspension bridges very much. The third was that the design should be suitable for the city. Of these three requirements, the first one was not difficult, because I thought I knew almost all the famous bridges in the world, and I could design a unique bridge for the city. The second requirement was also not difficult, since the tensile principle can be applied to a bridge in many ways, even if the bridge does not necessarily take the direct shape of a suspension bridge.

The third requirement, however, was very difficult, because the city of Beppu has only one special feature—its hot springs—so I was at a loss as to how to incorporate this feature in the design of a bridge. I visited the city several times, and at one point I observed workers renovating the pavement of a walkway. Blocks of very fine granite stone were being worked on at the site. I took some pieces of the stone to my laboratory and tested them. They were three times as strong as concrete in compression. I asked the workers where the stone came from and was told that it had come from China, because they had a sister city in China, called Yantai, from which they had been importing the best-quality stone. It occurred to me that if I used that stone for an important part of the bridge, it could serve as a way of promoting friendship between those two cities. I thought this might be a suitable contribution to the city of Beppu.

The results of my design can be seen in figure 1. The upper chord of the bridge is not concrete but granite stone blocks imported from China. For the lower chord I used a kind of chain system. Between the two elements, I put a series of lattice members that is isolated from the adjacent ones. This bridge looks different depending on the time of day and the direction of the sun. In the morning and evening, the lower chord can be seen very clearly, so there is an impression of the bridge as it actually is. But around noon, this chord is in the shadow of the deck, and the upper chord makes a stronger impression.

Because the lattice is not continuous, the upper deck bends to local loading, but the calculations proved that the stone blocks were fine under any and all stress conditions. The schematic drawing for the constitution of the bridge is shown in figure 2. We varied the lengths of the stone blocks, which are 25 centimeters deep and 40 centimeters wide, along the span of the bridge. These blocks were prestressed by means of five prestressing wire strands penetrating them. When the whole upper chord stone was prestressed, it became a kind of monolith. I tried to design the handrails to be as simple and beautiful as possible. At the joints of the lower members, I put ceramic ornaments (fig. 3), which are the only decoration on this bridge. This kind of ornament was not my invention but rather followed a traditional means of hiding nails that was used at the intersection of two timber members in old Japanese buildings.

Cast Steel Joints in the Space Frame of Expo '70

Steel structures as a structural system can be regarded as somewhat softer than stone structures. Of course, steel as a material is much harder than stone, but since we use only a small amount of steel in a structure, the resulting structural system can be said to be softer than a stone one. The example here is the Grand Roof for Expo '70 in Osaka (fig. 4), which was designed by Kenzo Tange. It was called a roof, but was not a mere roof. Its dimensions were 100 meters by almost 300 meters in plan. The depth of the roof, being 7.6 meters, allowed for a two-story exhibition space that accommodated roughly 5,000 visitors. This very large, heavy roof was supported by only six columns, which meant that the structural components of the space frame, the bars and the joints, had to bear extraordinary forces. The space frame was

4
Grand Roof of the Festival Plaza, Expo '70, Osaka

5
Detail of the cast-steel joints for the Grand Roof

6
View of the actual joint

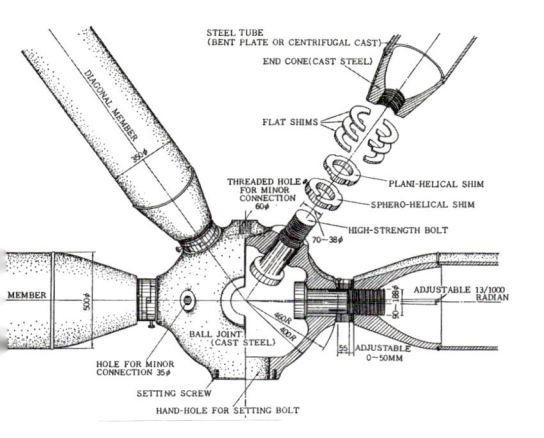

composed of a kind of mechanical joint of cast steel, which we designed specially for the structure (see figs. 5, 6). In the structure, we used cast steel extensively, although it was not the first time this kind of steel had been used for such a purpose. In 1964, at Yoyogi Indoor Stadium, which will be mentioned later, we adopted cast steel for the major components of a huge structure for the first time.

But in the Expo '70 structure, the use of cast steel was more explicitly expressed than in Yoyogi Stadium, and thus the stress flow of the roof was felt more directly. The late Peter Rice wrote in his autobiography, *An Engineer Imagines,* that his visit to the building after Expo '70 led him to the idea of adopting cast steel for his design of the Pompidou Center in Paris.

I think the construction method is very, very important in structural design. Some designers are enamored of the behavior of a structure only after it is completed, but I think the way in which a structure is constructed is very important as well in terms of rationality and economy. In the construction of this building, we applied what is called the lifting method, which was the first time in building structures such a method was used on a very large scale. The whole roof structure was

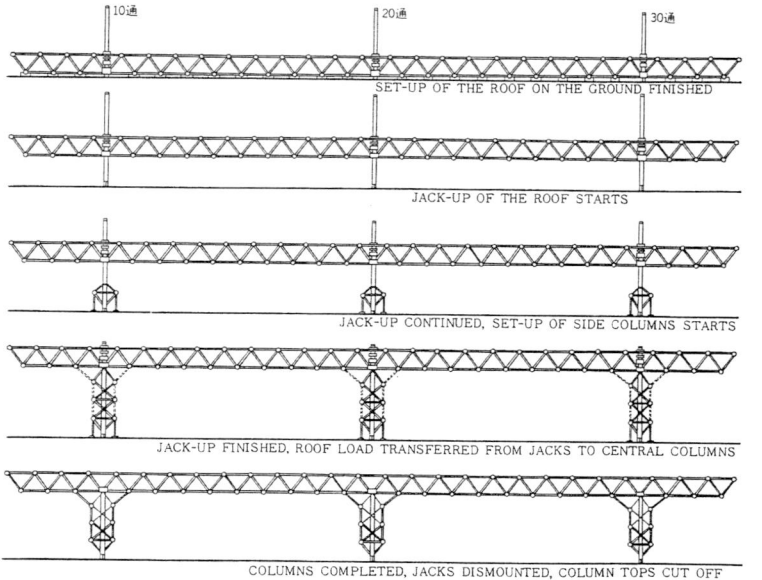

7
Lift-up scheme of
the Grand Roof

assembled at ground level and then lifted up by means of jacks—in this case, pneumatic jacks, not hydraulic—that climbed up along the six columns like monkeys (fig. 7).

Rational Erection of Domical Structures: The Pantadome System
Since the lifting of the big roof of Expo '70 was successful, it occurred to me that another kind of space frame that is not flat but in the shape of a dome might also be able to be lifted up using a similar principle. But as you can see in figure 8, it is not easy to fold a dome because it has resistance in the hoop direction as well as in the arching direction, and the combination of these two actions makes a dome very stiff and strong in comparison with an arch.

What I did was to remove the hoop members on a certain level temporarily so that the dome could be folded (fig. 9). The assembly work was done on the lower level, which facilitated safe and reliable construction. After every specified member was assembled on the folded shape, the structure was lifted by means of hydraulic jacks with temporary supports, or by means of blowing air inside. After the dome had assumed its final shape, those members that had been put aside temporarily were fixed to their specified positions. The structure was then complete. If you wanted to have your dome much lower in the first stage, you could put aside two rows of hoop members, making a double folding. The whole sequence of the assembly and erection process can be seen in the model (fig. 10).

I named this structural system the Pantadome system because its ruling mechanism is very similar to that of a pantograph, a drawing

106 | THE FELIX CANDELA LECTURES

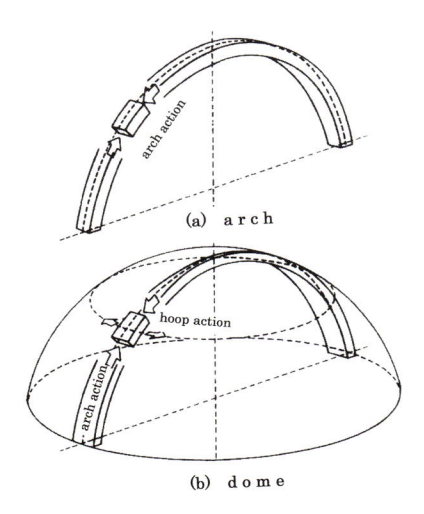

⑧
Sketch showing how a dome is less collapsible than an arch

⑩
Model study of the Pantadome system

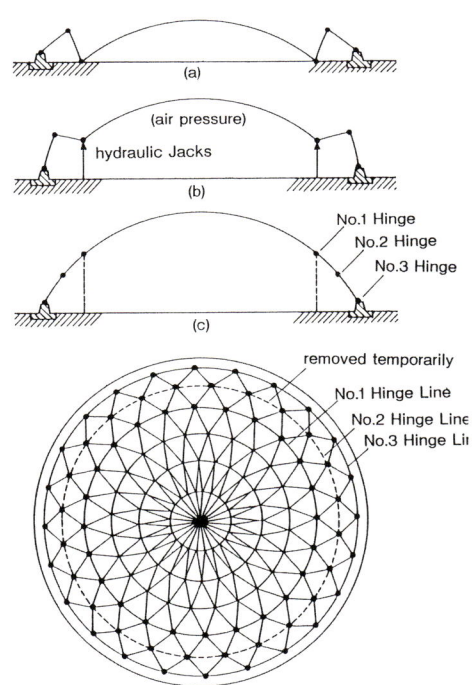

⑨
Principle of the Pantadome system

KAWAGUCHI | 107

NAME	WORLD	SINGAPORE	ST. JORDI	FUKUI	NAMI HAYA	NARA HALL	COAL STORAGE
SHAPE AND DIMENSION (SMALL CIRCLES IN PLANS INDICATE PUSH-UP POSTS)	110 m	200 m	128 m	116 m	127 m	127 m	251 m
BUILT	1984	1989	1990	1995	1996	1998	2001
COVERD AREA	7,700 m²	14,000 m²	12,000 m²	10,500 m²	11,000 m²	6,500 m²	40,000 m²
TOTAL WEIGHT	1,680 t	2,600 t	3,000 t	5,430 t	4,690 t	4,660 t	7,500 t
STEEL WEIGHT	760 t	1,250 t	950 t	2,770 t	1,160 t	—	6,500 t
LIFTING HEIGHT	20 m	20 m	32 m	28 m	29 m	14 m	30 m
LIFTING POINTS	18	12	12	8	16	32	14
SPECIAL FEATURES	OVAL PLAN FIRST ATTEMPT	RHOMBIC PLAN ABROAD	UNFINISHED SHAPE ABROAD	PURE CIRCLE HEAVY SNOW	INCLINED ROOF QUICK LIFT	PRESTRESSED CONCRETE UNITS	BIG COVERED AREA LIGHT WEIGHT

11
Realized Pantadomes. The Pantadome system has been applied to several major spatial structures of different shapes and dimensions.

instrument used to produce similar figures until a few decades ago. This system has been successfully applied to several buildings, as you can see in the table (fig. 11). One of the most important features of the system is that during the lifting stage the dome follows a mechanism that has only one degree of freedom in movement. This means that the dome moves without resistance only in one direction, while exerting a great resistance against movement in all other directions. So you don't have to worry about lateral forces due to wind or earthquake, which may attack the structure during its erection. Doing without bracing or guying the structure in its construction enables you to concentrate on the lifting.

World Memorial Hall (1984) was, in plan, 110 meters by 70 meters. This hall happened to have been built very close to the epicenter of the Kobe earthquake that occurred eleven years later, and while it experienced severe tremors, it did not suffer any damage. The second application of the Pantadome system was Singapore National Stadium, which was designed by Kenzo Tange in 1989. Its dimensions are 200 meters in the longitudinal direction and 128 meters in the transverse direction. It has a section that is closer in shape to that of a hanging roof than to a dome. The third one, designed by Arata Isozaki, was the Sant Jordi Olympic Sports Palace in Barcelona, which was built for the Barcelona Olympics of 1992. We also applied the same system to other structures, as the table shows.

Now let me mention a few examples of the Pantadome system in some detail. In the Sant Jordi Olympic Sports Palace, hinges were placed at the bottom and top of the columns and at intermediate points of the roof. There are a total of six hinges in one section. If this were a plane structure, such a structural system would be prohibitive since it would be highly unstable. But if you think about the three-dimensional behavior of the structure, you will see that it is very stable. Figure 12 shows the construction process of the building, while the exterior and interior views of the completed structure are shown in figures 13 and 14. Isozaki wanted to have the shape of a dome that was not a complete dome, but one that suggests the way the dome had been constructed. He wanted a shape that seemed as if the dome was on the way to being lifted. In the same sense he left those parts that moved very much during the construction as they were, and those spaces were used as skylights (see fig. 14).

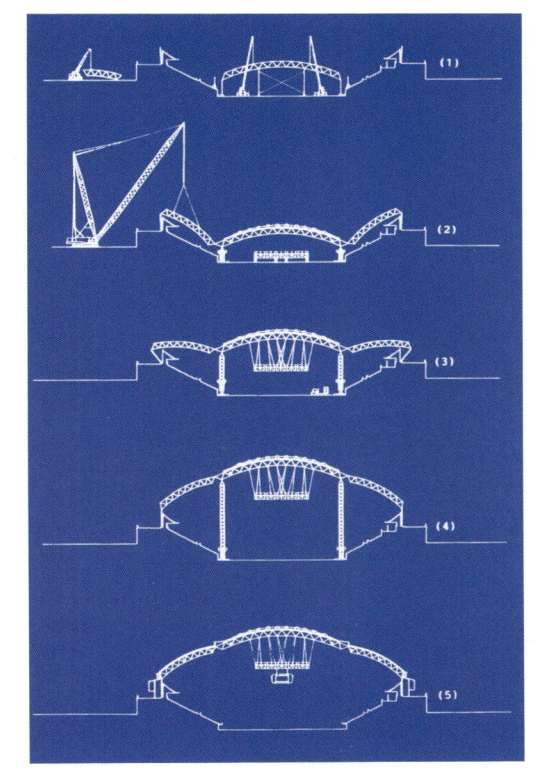

12 Erection process of Sant Jordi Olympic Sports Palace

13 Exterior view of Sant Jordi Olympic Sports Palace

14 Interior view of Sant Jordi Olympic Sports Palace

KAWAGUCHI | 109

15 Lifting of Namihaya Dome, which was completed in 8 1/2 hours

The next example of the Pantadome system is Namihaya Dome, a swimming pool constructed in Osaka in 1996 (fig. 15). It has an oval plan of 120 meters by 110 meters. In elevation the equator of the surface is not horizontal, but inclined by five degrees. So the axis of the surface is also inclined five degrees from the vertical. After the assembly work was done, the roof was lifted, not vertically but in an inclined direction. In general, it is prohibitively dangerous to lift a half-built structure in a direction that is not vertical. But as you saw in the previous discussion, a Pantadome structure has a freedom of movement only in the specified direction, in this case along the axis of the dome that is five degrees inclined from the vertical. The structure is resistant in any other direction by itself. The erection sequence of the dome is shown in figure 15.

The roof could be lifted in a non-vertical direction without any problem. We did not have to worry about any lateral forces during the erec-

110 | THE FELIX CANDELA LECTURES

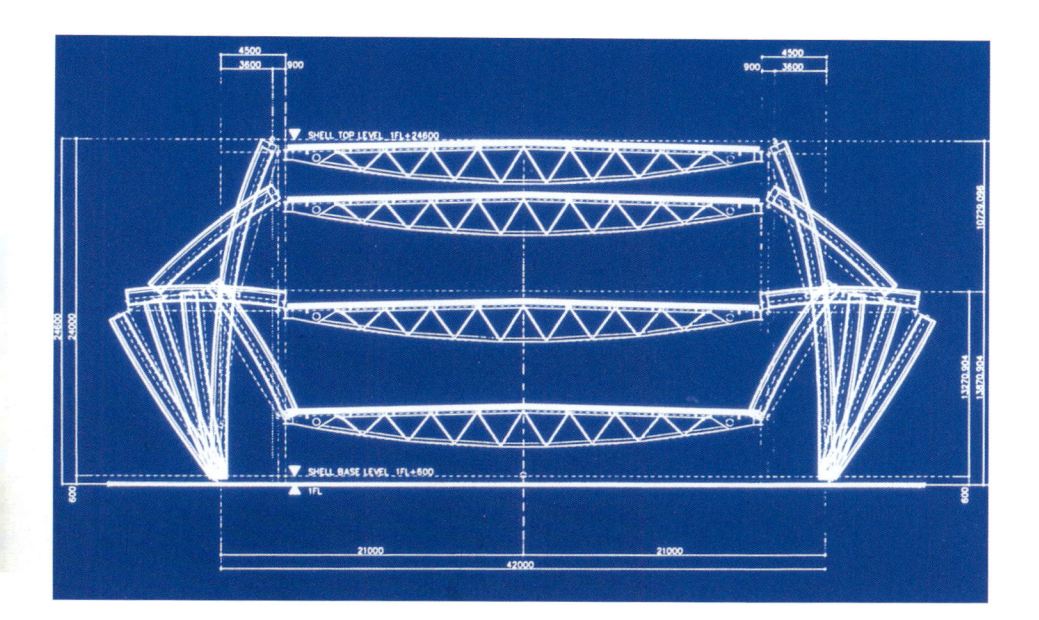

16
Lifting scheme of Nara Centennial Hall, Nara

17
A view of Nara Centennial Hall during erection

tion. Actually while this dome was being constructed, it experienced strong seismic motion due to the Kobe earthquake, but it suffered no damage or inconvenience in construction. The whole roof was lifted in eight and a half hours.

The last example of the Pantadome system is Nara Centennial Hall, designed by Isozaki and completed in 1998. In this case we folded not only the roof, but also the wall. After assembly, the dome was lifted to its final position. The structure took various shapes during the lifting process according to the position of the roof (see fig. 16). The shape of the building was much more dynamic in the process of construction than it was after completion (fig. 17). It has since occurred to us that it might be interesting if we could have a structure, or building, that could move in this way even after completion and would not pose any problems in terms of technology. The only difficulty would be building codes.

KAWAGUCHI | 111

Soft Steel Structures

A much more flexible and softer steel structure is a suspension structure. An example is the Yoyogi Indoor Stadium (fig. 18), designed by Kenzo Tange and Associates for the Tokyo Olympics of 1964. The structure was by Yoshikatsu Tsuboi and his institute. I was the chief engineer of Tsuboi Institute at that time, and was in charge of the structural design of the major building. Before the design began, we worked closely with the architects' group, and we made many small models in the same room. I think that the architects' and engineers' cooperation at this stage was a very important factor in enhancing the design value of the building.

Figure 19 shows the interior of the stadium, which is aesthetically pleasing as well as excellent in acoustics. The structural system can be seen in figure 20: the central structure is similar to a suspension bridge. When we began our design, we discovered that we had no reliable basis on which to calculate our roof structure. So we had to develop the fundamental equations for those parts that are network structures with some bending stiffness, which we called "semirigid" hanging roofs. We did not have any calculating machine that could be

18 Yoyogi Indoor Stadiums for the Tokyo Olympics. Stadium I is on the right.

19 Interior view of Yoyogi Indoor Stadium I

112 | THE FELIX CANDELA LECTURES

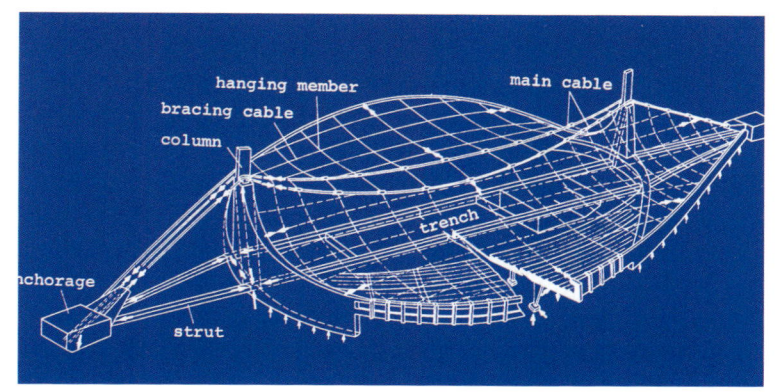

⑳ Schematic structural system of Stadium 1

㉑ Cast steel Saturn joint for Stadium 1

㉒ Saturn joint in position

called a computer, so we had to use a hand-driven adding machine for the calculations. Nonetheless, we were able to obtain a safe and sound result with sufficient accuracy to complete our structural design.

The next difficult issue in design was the detailing of the structure. One of the difficulties stemmed from the high flexibility of the roof. In fact, the whole structure was very flexible, so flexible that the central part of the main cables displaced downward as much as 2 meters during construction. We had to design every detail to accommodate this big deformation or displacement. For the connection between the main cables and the hanging steel members, we developed a detail that is very similar to the planet Saturn with its satellite ring around it (figs. 21 and 22). This central sphere of cast steel catches the main cable, and the ring can rotate around the sphere. This detail proved very useful for absorbing the big deformation of the main cable during construction. Another advantage of this design was that the same joint could be used for every connecting point.

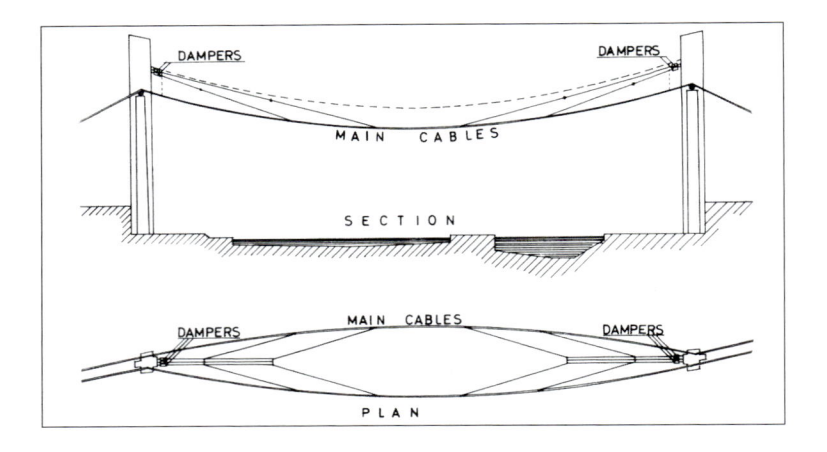

② Damping system for the roof

② Oil dampers visible from outside

Another interesting feature of the building was a damping system, which may control possible wind vibration. Because the roof was very light, there was no problem of earthquake resistance, but we had to be careful about the dynamic effect of wind. We carried out a wind-tunnel test and all the necessary dynamic calculations, and we became confident that no danger would happen to this building due to strong wind. But the client wanted the building to be as strong as possible so that it could be used as a kind of disaster-prevention center in case of destructive situations. So we installed a group of oil dampers on the main cables of the monitor roof at both ends (fig. 23). Tange designed these dampers to be visible from outside (fig. 24). They are actually a series of dampers but at the same time they were used as a small piece of architectural expression. To my knowledge this was the first application in the world of a damping system to a structure.

Soft Structures: Membrane Envelopes

Returning to the Expo '70 space frame (fig. 4), I would like to mention a very soft structure, a transparent roofing system set on top of the gigantic space frame. The top surface of this frame consisted of 274 grids that were 10 meters by 10 meters. On top of each of those grids we designed a transparent pneumatic panel (fig. 25). The top and bottom skins of this panel consist of layers of very thin, transparent polyester film that was extended in biaxial directions. The total thickness of the skin is only 1.2 millimeters. Figure 26 shows the full-scale model of the transparent roof panel for tests. At first nobody wanted to step on the roof because it seemed very weak, but after someone stepped on it timidly and looked safe, the workers wanted to climb on the model one after another, and finally they began to jump on the roof, enjoying this kind of trampoline.

There is still a softer structure that is made of fabric. One kind is air-supported, the other air-inflated. As you can guess, the air pressure necessary in an air-supported structure is very low (10 to 30 millimeters Aq.), while in an air-inflated structure, the air pressure needed is very high (100 to 1,000 millimeters Aq.) compared to the former.

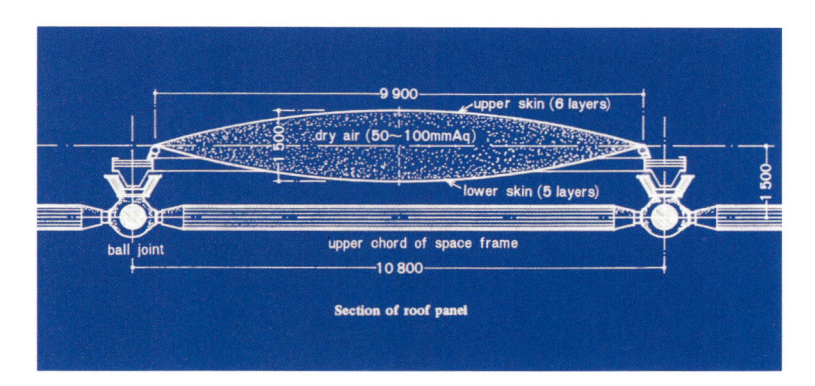

25 Transparent air panel for the Grand Roof of Expo '70

26 Full-scale test model of the air panel

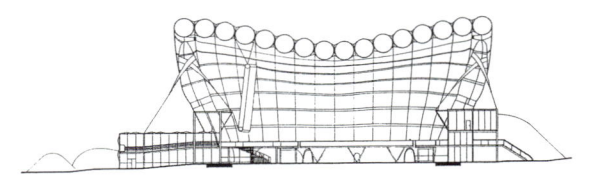

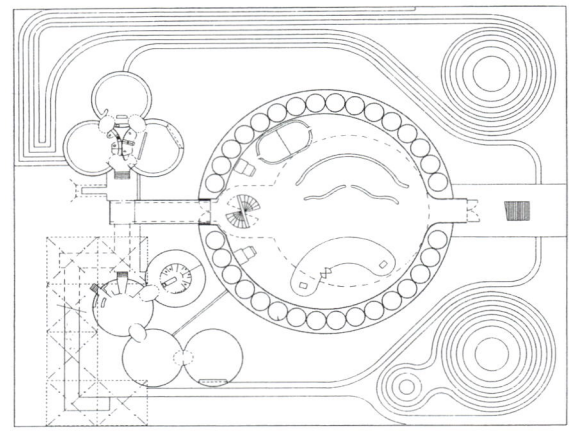

27 Plan and section of Fuji Group Pavilion for Expo '70

28 Fuji Group Pavilion, exterior view

    The Fuji Group Pavilion (figs. 27 and 28) at Expo '70, which I worked on with the architect Yutaka Murata, is an example of an air-inflated structure. It is circular in plan, 50 meters in diameter, and in shape possesses a very clear and interesting geometry. The whole structure consists of sixteen arches of the same length. The two arches at the center have a semicircular profile, but the distance between the two footings of an arch gets closer as the location of the arch shifts from the center toward the ends. Since the total length of the arch is constant, the top of the arch becomes higher from the center toward the ends of the pavilion. This principle affords a shape to the building that is unique and perfectly defined by geometry. The air pressure inside the tubes was kept about 1,000 millimeters Aq.

    Murata and I also designed an interesting air structure for the Electric Power Pavilion Annex in the same exposition (figs 29 and 30). It

㉙
Electric Power Pavilion Annex for Expo '70, exterior view

㉚
Electric Power Pavilion Annex, structural system

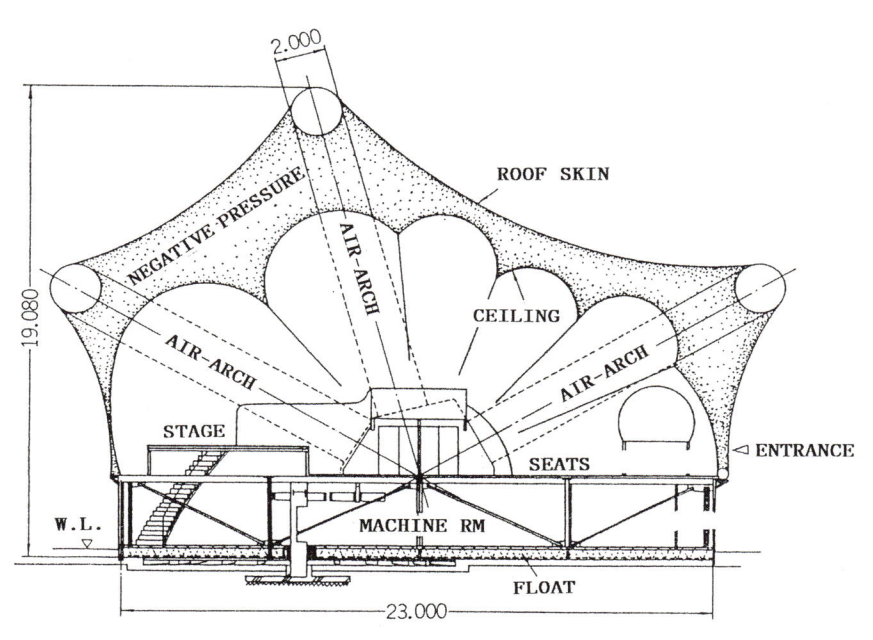

was a theater for magic shows, and its roof consisted of three air-inflated arches supporting a membrane roof. Between the roof and the ceiling there was a space that was brought to negative air pressure. This means that the air pressure in the space between the roof and the ceiling is lower than the inside and outside air pressures, which are both atmospheric. The roof was then pressed inward due to the differential air pressure, while the ceiling was lifted upward by the same pressure difference, the ceiling doing without any structural support to keep it in position.

The theater was actually a boat floating on a pond and was positioned in such a way that while the audience was enjoying magic shows inside, the theater was turning slowly around a pivot. After the shows ended, people found themselves disembarking at a "port" that was different from the one where they had embarked, thus enjoying

KAWAGUCHI | 117

another kind of magic! The theater rested on a specially designed float that did not have a "bottom," but had only a "ceiling." The space between the ceiling and the water was pressurized to produce buoyancy, and the boat was stabilized by several bulkheads.

Mesh-Reinforced Air Structures
The skin of pneumatic structures usually plays a dual role: maintaining the differential air pressure between exterior and interior spaces and resisting the tension induced in it by the external loads as well as by the air pressure. For larger structures, in which the fabric skins are not strong enough to span the whole roofing areas, cables are sometimes employed to reinforce the skins. In this case, the skins act more as air tighteners than as structural tension members. The result is a rational pneumatic structure in which the above dual role is separately taken by different systems: skin materials for air tightness and mesh, and networks and/or cables for tensile resistance.

Mesh-reinforced air domes jointly developed by Murata and me are examples of such pneumatic structures. The skin of the domes consists of a very thin plastic film pressed by the differential pressure against a net of comparatively fine meshes that occupies the space between boundaries and/or reinforcing cables. The role of the plastic film is to form a transparent surface to keep air pressure inside. And as the size of the mesh is on the order of a few centimeters, the tensile strength required for the film is very small. Other properties of the film, such as airtightness, ductility, transparency, and durability, are therefore more important. No connection is needed between the net and the film, as the latter is in close contact with the former by the differential pressure. For an area where a high external pressure is expected to occur under strong wind, another layer of the net and film is overlaid to prevent the inner film from being pushed apart from the

31
Mesh-reinforced air structure for play yard

32
Twelfth World Orchid Conference Pavilions, Tokyo

net. One of the special features of this system is that there is no concern about the cutting patterns of the net or the film, and domes of different boundary shapes can be produced by a flat net and a flat film.

We designed this type of air structure for agricultural and fishing purposes. In these temporary buildings, everything had to be inexpensive. Polyvinyl chloride (PVC), used for the enveloping film, was reinforced by a fishing net that was also very cheap. Wire rope was used only for important parts of the structure. Some of these structures were used in hydroponic agriculture. A similar system was applied to cover a pond in which eels were cultivated. By covering the pond with a transparent membrane, the temperature of the pond water was kept warm and the eels grew very quickly, even in wintertime. In one of our designs, the same system was used for a play yard because the film, inflated by air, was soft and safe and exciting for children to play on (fig. 31).

A more recent example of the same system is the two pavilions (fig. 32) for the Twelfth World Orchid Conference held in Tokyo in 1987 and designed by me in cooperation with the architect Murata. Dome 1 of the Pavilion is a circular plan, 75 meters in diameter, with a height of 19.5 meters. The air dome is reinforced by two-way wire ropes spaced at intervals of 5 meters. Each bay of the dome, bounded by wire ropes, is structurally constituted by fishing net of 10 centimeter mesh with PVC film of 0.1 millimeters thickness for airtightness located inside the net. The internal air pressure is kept at 30 millimeters Aq. above the atmosphere for normal conditions, and is increased to 70 millimeters Aq. when there are strong winds or heavy snow. Dome 2 is shaped like a worm that is 40 meters wide and 100 meters long. The dome is reinforced by wire ropes mainly in the transverse and radial directions, with the longitudinal reinforcement only along the ridge. The constitution of the skin is the same as in Dome 1.

Softest and Lightest Structure: Fabric in the Natural Air
There is a tradition in Japan for people to celebrate Children's Day, May 5, by flying carps made of cotton fabric. These carps have almost exclusively been made in the small city of Kazo, which is located in a suburb of Tokyo. The normal size of flying carps is 3 to 5 meters in length. Some years ago, I was consulted by the city about the possibility of flying a 100-meter-long carp that residents had made for the purpose of advertising their city. There were three questions asked: Could the jumbo carp fly? If it could fly, how would it fly? Could the cotton skin stand stress during the flight?

The most interesting aspect of the problem was that the jumbo carp had been made from the same fabric as carps of normal size. A

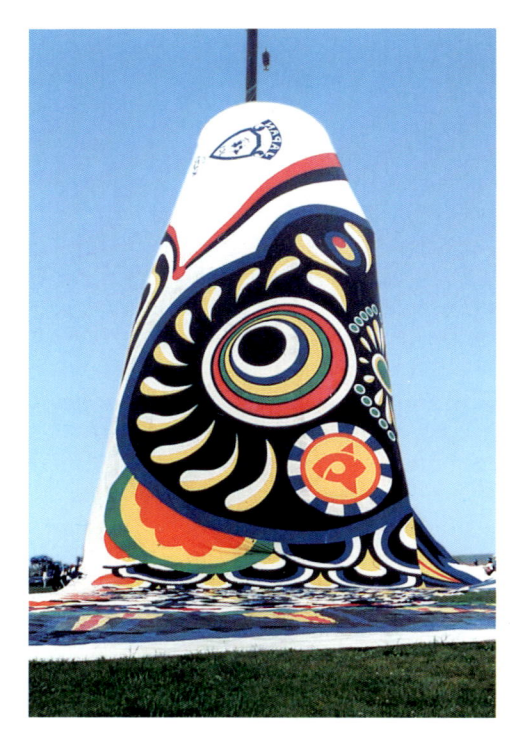

33 and **34**
100-meter-long fabric carp being lifted by a crane in a breeze and soaring in the sky

dimensional analysis for flying conditions of the jumbo carp, a calculation of membrane stresses, and a series of wind-tunnel tests were undertaken. They revealed that the jumbo carp flies at the same wind speed as a carp of normal size, and it swims very slowly compared to a normal one. The skin is subject to twenty to thirty times higher stress than that of a normal-size carp during flight. The cotton fabric is strong enough to stand this stress, but the seams of the fabric are too weak. Other technical problems included such details as the mouthpiece of the carp and how to raise the carp to a reasonable height.

On the basis of the above information, the jumbo carp was reinforced along the seams, provided with a specially designed mouthpiece of aluminum tubing, and raised by means of a truck crane on a fine day in April 1988 (fig. 33). The jumbo carp, a huge flying membrane, began to fly with the breeze and swam elegantly in the sky (fig. 34). Since that time, the flight of the jumbo carp has been one of the most important annual events of the city, celebrated at the beginning of May.

❶
Joseph Strauss. Golden Gate Bridge, San Francisco

The Art of Bridge Design
CHRISTIAN MENN

Structural design falls between two opposite poles: natural science and art. Natural science is static, exact, and indisputable; everything can be precisely measured and checked, and artistic freedom does not exist. In contrast, art is dynamic, open, creative, and disputable; nothing can be measured or checked, and artistic freedom is unlimited. Genuine bridge design, which consists of at least ninety percent applied mechanics, is much closer to natural science than to art. Accordingly, artistic freedom in bridge design is limited to physical laws, and the form of a bridge is largely related to clear, rational, and aesthetic criteria. Buildings, by contrast, are more closely allied with art, and one's appraisal of their architectural expression can hardly be made on a rational basis.

Genetics of Structural Engineering
There have always been engineers, if we understand the term to mean those who calculate measurements prior to the erection of a structure, taking into consideration functionality, safety, and durability. The construction of the pyramids of ancient Egypt, which involved the production, transportation, and erection of more than two million stone blocks, was a typical engineering problem. In ancient times it was not possible to achieve material prosperity and build a state system without the support of engineers. During the Roman Empire, for example, roads, bridges, city fortifications, canals, and aqueducts were of vital importance.

However, in ancient times, as in the Middle Ages, the engineer was generally not a specialist. He worked with the architect and often with the craftsman, and was likened to a master builder. But it was always the engineer in his capacity as master builder who made possible the success of the architect and the craftsman, not the other way around. The Roman architect, engineer, and writer Vitruvius, in his *Ten Books of Architecture*, stated that structural safety was the most important goal of architecture.

The role of the engineer as a master builder can be seen in Byzantine architecture, such as in the fantastic structure of Hagia Sophia, built in

Constantinople in the sixth century. The engineer continued to be important in the building of European cathedrals, domes, and spires of the Renaissance and Baroque periods. The better the engineer as master builder, the bolder these magnificent structures were. Filippo Brunelleschi, the famous master builder of the Cathedral of Santa Maria dei Fiori (Duomo) in Florence, was not only a great architect but also an excellent engineer who could clearly visualize the flow of forces in a dome structure. His vision enabled him to succeed in constructing the 42-meter span of the dome of the cathedral without any scaffolding.

During antiquity and the Middle Ages, supporting, or bearing elements were not only in almost all major public and religious buildings and monuments but also in bridges and aqueducts that blended into a harmonious culture of building. Leon Battista Alberti, a leading architectural critic during the Renaissance who was inspired by Vitruvius, praised the art of building because it comprises mathematics, technology, and art. In those days, structural engineering and architecture were closely integrated and balanced between mathematics and natural science on the one hand and art on the other.

Toward the end of the eighteenth century, worldwide political revolutions and radical industrial changes were accompanied by extraordinary changes in the building industry. Paving the way for this revolution were the great mathematicians and scientists of the seventeenth and eighteenth century—Isaac Newton, the Bernoulli family, Gottfried Wilhelm Leibniz, Leonhard Euler, Charles-Auguste de Coulomb, and Carl Friedrich Gauss, among them. Their writings and theories encouraged the nineteenth-century French civil engineer, Louis Navier, to apply the principles of mechanics to analyze structures for the purpose of ensuring adequate structural safety using the smallest amount of material.

At this time, the engineer and the architect went their separate ways. The engineer, whose mandate was to guarantee an acceptable level of safety with a minimum amount of material, increasingly turned to natural science. The architect, freed from the issues of structural safety, turned more and more toward art. This trend led to the disruption of the integrated objective of the art of building, which consists basically—besides impeccable structural safety—of an optimal synthesis of functionality, aesthetics, and economy.

Today, renowned architects try to achieve a balance between functionality and aesthetics, paying less attention to economic considerations and structural issues. In contrast, engineers regard the satisfaction of structural safety requirements at the least possible cost as the most important sign of quality, and they view any functional deficiency as a failure. They leave all problems of careful shaping to "others," like architects, politicians, interest groups, etc. This dissimilar perception of the fundamental objectives of the art of building led to an unpleasant estrangement and polarization between architects and engineers.

Intensive research has been conducted in structural engineering during the last fifty years. As a result, the engineer's profession has become more and more centered on science. In the last few decades, his training has consisted almost exclusively of scientific-analytical subjects; engineering standards have been greatly expanded, while the creative side of structural engineering has continued to decline. This new one-sided science-oriented development of our profession is, however, not convincing. First of all, natural science is only one pillar of our profession. Research has but a marginal influence on structural engineering compared to other fields such as mechanical engineering, the automotive and aircraft industries, electronics, biology, pharmacology, medicine, etc. The Golden Gate Bridge in San Francisco (1937; fig. 1), for example, although built some seventy years ago, would still be considered a convincing project today. In fact, many recent bridges seem hardly to have progressed beyond those built in the 1930s, at least from the point of view of the layman. The same is not true in other fields, where the products of today differ enormously from those of the 1930s.

Like architecture, but perhaps to a lesser extent, structural engineering requires practical experience, innovative ideas, imagination, fantasy, and art. This is the second pillar of our profession, which is important and has nothing to do with analytical investigations, although in structural engineering, creativity and art cannot be separated from natural science and must build on knowledge gained through scientific research.

Consequences of the Scientific Emphasis of Structural Engineering
Today, like two hundred years ago, we are in the midst of an economic revolution that is accompanied by great political changes. This time the economic revolution has been triggered by information technology and globalization, and in this process the scientific aspect of our profession is producing dubious results. The following trends have become particularly noticeable in structural engineering: the overrating of analysis and underrating of creative work; the standardization of bridge engineering; and the strong impact of outside influences on engineers.

Overrating of analysis and underrating of creative work
Structural analysis nowadays can be performed with ever-increasing accuracy due to the availability of computers. Structural standards are undoubtedly encouraging this development, and many engineers, and to a greater extent their clients, seem to fully believe that the quality of a structure is related to the amount of computation involved in its creation. As a result, unfortunately, more and more engineers are computing too much, too early, and are thinking too little, too late.

In contrast to this, the creative work of an engineer is little appreciated. With good ideas one can, at best, but certainly not always, win a design competition. But, in any case, I have never met a client who would be willing to honor an interesting conceptual design. In this respect, we have clearly moved far away from architecture and art, where unusual ideas are mostly acknowledged and rewarded with prizes.

Standardization of bridge engineering
Excessive codification has inhibited creativity because of restrictive building clauses; even for a standard project, many engineers are understandably overwhelmed by the need to satisfy numerous and complex clauses. Therefore, many of them tend to exclude unconventional structural solutions from the very beginning.

Strong impact of outside influences on engineers
The most worrisome aspect of the scientific emphasis of structural engineering is the increasing impact of outside influences on engineers. They have less and less freedom to make decisions, which are more and more confined to craft-computational aspects only. Our profession had its time and chance, as Le Corbusier said: "The engineer, inspired by the laws of economy and directed by calculations, puts us in harmony with the laws of the universe." Since that time, the glamour of our profession has more and more faded away.

The present tendency toward banal bridge engineering, even in prestigious projects, has led to the placing of architects, often so-called star architects, above bridge engineers. These architects, who consider bridge design only as a hobby or a market slot, are convinced that they can design bridges even though they have no structural knowledge. Certainly authorities, the public, and, unfortunately, even engineers themselves believe that. Today there are those who submit bridge designs by picture book. Such proposals are sometimes judged by an incompetent jury, and awards are granted despite the fact that structural principles or costs have not been taken into consideration. Such a design procedure would be completely unacceptable in architecture or in the fine arts.

I believe that our profession is now at a crisis, and if we want to emerge from it and make our profession attractive again, we should, first of all, change the education of engineering students. Globalization has made business professions more attractive, and interest in the civil-engineering profession is already on the decline. Moreover, the younger generation with scientific talents is drawn more toward high-tech professions than to civil engineering, while those with creative and artistic skills would rather choose architecture because they do not see any room for creativity in civil-engineering education and practice. Hence, the fundamental policy of engineering education should be to promote and strengthen the creative pillar of our profession.

The Goal of the Art of Bridge Design
The goal of the art of bridge design consists, in my view, of developing a structural system that achieves an optimal balance between economy and aesthetics, keeping in mind the prominence of the bridge

site. First, the functional demands—the type of traffic, roadway alignment, and carriage cross sections—must be fulfilled, as must structural safety, serviceability, durability, etc. Second, the technical and environmental boundary conditions concerning topography, geology, hydrology, seismology, clearances, construction duration, local construction technology, landscape and surrounding constructions, water and vegetation, and emissions must be satisfied.

In bridge design, aesthetics stands, on the one hand, for perfect integration of the bridge with its environment and, on the other, for a rational and artistic refinement of the raw structural form. The prominence of the bridge site means, above all, the visual exposition and the prestige of the bridge in its surroundings. In practice, a raw conceptual design is first developed that satisfies the above-mentioned demands in such a way that the bridge form fits perfectly into the environment in space and time. To fit in space means to be suitable in scale and optimal in form to the topography and the clearance requirements; to fit in time relates not only to the state-of-the-art of structural engineering but also to local history and tradition. However, history and tradition do not mean that a historic style should be copied; rather, it means that the historic style should reappear in a modern technological form.

The conceptual design is always based on perfect structural engineering, which leads to original and logical forms in a natural manner, as seen, for example, in the Sunniberg Bridge near Klosters, the Charles River Bridge in Boston, or in the bridge with a free span of 3,000 meters that I once proposed. It is only the skilled and experienced structural engineer who can develop a genuine bridge concept.

After the conceptual design of a bridge has been developed, the raw structural concept is refined. Since bridge engineering is close to natural science, a number of clear and rational criteria, which can be measured and checked, should be used. These include: visualization of technical efficiency through slenderness, transparency, and tenseness; order and unity of the entire structure through clear organization of the structural system and its components and a coherent, unified typology of the cross sections of all structural components; visualization of the flow of forces through adequate development of the cross sections of the components; and artistic shaping through clean detailing, light and shadow effects, structural ornamentation (evolving effects, e.g., arrangement of stay cables), and nonstructural ornamentation (colors, lighting, etc.). The conceptual design and its subsequent refinement must always be made keeping in mind the economic aspects and the prominence of the bridge site, which means an important balance between economy and aesthetics.

It is generally not difficult to estimate the cost of the most economical solution for a structure. In my opinion, the maximum extra costs should be, for medium-span bridges, no more than fifteen to twenty percent, and, for long-span bridges, within three to five percent of the cost of the most economical solution. Higher extra costs indicate that the basic concept was wrong, and one should start all over again. The real, but certainly never quite achievable, art of bridge design would consist of the highest aesthetic quality at the lowest possible cost.

Illustrative Examples
The following examples illustrate my thoughts about genuine bridge design.

Golden Gate Bridge in San Francisco
The bold concept of the impressive Golden Gate Bridge, completed in 1937, corresponds in space to the scale of the vast landscape reaching into the Pacific Ocean and in time to the highest level of bridge design and construction technology at that date (fig. 1). The longest span in the world when it was built, it symbolized a breathtaking gateway between San Francisco Bay and the Pacific Ocean. The integration of the bridge with its environment is, in space and time, simply perfect.

The shaping of the bridge is interesting as well as somewhat controversial. Spanning the bay with two pylons and a slender girder represented an extraordinary technical feat. A satisfying structural uniformity has been achieved in the main span, despite the fact that the truss typology of the pylons and of the bridge girder is different. The pylons, designed by the architect Irving F. Morrow, are of a decorative shape that mirrors the flow of forces in a fascinating manner. The one to two percent additional cost that was incurred because of the special ornamentation of the pylons was absolutely justified in the eyes of those who use the bridge. However, the choice of structural forms for the approach viaducts, particularly on the San Francisco side, was neglected in the overall shaping. The arch over Fort Point is an alien element; it disturbed me even in my student days. Today, however, one could characterize it as a sort of "piercing," or a unique feature of the Golden Gate Bridge.

Joseph Strauss, who designed the bridge, was certainly fascinated by the design, but one feels that he was not especially captivated by the form of his bridge if one looks at the proposal he put forward in 1919. I suppose that it was a unique, lucky coincidence for the design process of the Golden Gate Bridge that in 1923 Othmar Ammann published his famous "Study of a Highway Bridge across the Hudson River at New York, between Washington Heights and Fort Lee," in which he proposed a very long span that would reduce the depth of the stiffening girder instead of enlarging it. The greatness of the Golden Gate Bridge, however, compensates for the relatively minor defects in its uniformity. From afar, these aesthetic shortcomings are hardly noticeable, and, up close, each section of the bridge appears as a distinct, individual structure.

Salginatobel Bridge near Schiers
The Salginatobel Bridge of 1930, which spans a canyon, is situated at a very favorable topographical position, and its select structural concept

Salginatobel Bridge, Schiers, Switzerland

perfectly suits the local topography (fig. 2). The flat arch spanning the deep valley undoubtedly offers a feeling of an extremely daring and efficient structural system. The arch and the approach section have the same structural forms. The shape of the arch clearly depicts the flow of forces. All cross sections exhibit the same typology: arch, piers, and bridge deck are all of the same statically efficient T-beam type. Artistic ornamentation, reduced to a minimum, has been used only in the abutment areas for the visualization of both the flow of forces and for structural stability. With the Salginatobel Bridge, the structural engineer Robert Maillart attained practically the ideal goal of structural art in bridge engineering: that is, he obtained the most pleasing aesthetic quality at a minimum cost. Therefore, even if it were designed now, some seventy years after its construction, the Salginatobel Bridge would still be a truly valuable project.

I would now like to review a few of my own bridges and discuss them critically, revealing some lessons that I learned regarding the art of bridge design.

Bridge over the Averserrhein at Cresta
This was one of my first bridges (fig. 3). It was designed in 1960 in part as an homage to Maillart. This arch bridge has been reduced to its basic elements: the arch and the deck girder, without any columns. With some minor technical improvements, this project would be economical even today.

3
Averserrhein Bridge,
Cresta, Switzerland

4
Averserrhein Bridge,
Cröt, Switzerland

130 | THE FELIX CANDELA LECTURES

Rhine Bridge, Reichenau, Germany

Bridge over the Averserrhein at Cröt

The deck-stiffened hinged arches of this bridge required a very light scaffolding for the arch (fig. 4). The secondary deck spans had to be small in view of the buckling of the slender arch. In 1960 this type of arch bridge was acceptable, but today the vast amount of formwork for the slender columns would be too expensive, especially for arches with a relatively high rise.

Bridge over the Rhine at Reichenau

This elegant arch bridge, completed in 1963, has a slender, transparent appearance due to the complementary relationship between the arch and the deck as well as between the approach spans and the arch span (fig. 5). It also displays a remarkable technical efficiency. All structural components are well organized, the spans are carefully staggered, and the arch, columns, and deck, which have the same simple rectangular cross sections, reveal a perfect order and unity of the entire structural system. The additional costs, when compared to an arch system with the most economic cross sections for each structural component, would have been approximately five to seven percent, which was absolutely justified in view of the spectacular bridge site. I built a number of similar arch bridges between 1961 and 1967, all of which integrate very well with their surroundings. Today the construction of these classical arch bridges is too expensive, but I think that we now have this possibility of building new, interesting arch systems.

Felsenau Bridge in Bern
The 1,100-meter-long Felsenau Bridge was built between 1972 and 1974 (fig. 6). It spans the Aare River in Bern, the capital of Switzerland, and therefore assumes a major prominence. But in spite of this prominence, some small financial concessions were made. These included using relatively big spans and narrow piers to achieve better transparency. In retrospect, I think that it would have been perfectly justifiable to spend five percent more in order to achieve a lighter, slimmer deck girder and more elegant, artistically shaped piers.

Ganter Bridge near Simplon Pass
This bridge, which began with a conceptual design in 1974, was completed six years later (fig. 7). In no other bridge did I learn as much as I did here. First of all, it became clear to me that any proposal that cost thirty percent more than the most economic solution is very likely to have some conceptual weakness and cannot be satisfactory. I also realized that structural elements above the roadway are always critical: technically, because they are exposed to rainwater and melted saltwater splashed by vehicles; and visually, because these structures appear to users riding over the bridge in a shortened view, and from this visual angle, in particular, concrete structures look rather heavy and clumsy. Unfortunately, we did not realize this effect on the bridge because the model on which it was based was probably too small.

Biaschina Bridge near Giornico
For this bridge, which was completed in 1984, I prepared only the conceptual design for the design competition (fig. 8). The bridge site is very impressive and special because the bridge crosses the Leventina Valley at a height of 100 meters and can be seen from completely different levels and angles from the famous Gotthard railway, which winds through three spiral tunnels at this location. The construction cost was only a small percentage above the minimum possible cost. Here, also, in view of the highly visible bridge location, the slenderness and elegance of the piers could have been further enhanced at a slightly higher cost.

Chandoline Bridge at Sion
This cable-stayed bridge has been in operation since 1987 (fig. 9). It crosses the Rhone in a slight curve at an extremely skewed angle and at very low height. Once again I realized the importance of sticking to certain simple shape criteria; for example, the piers and the pylons above them should basically have the same cross-sectional form. Moreover, the cables should be so arranged that a structural orderliness is evident from all viewing angles. I feel that many cable-stayed bridges have tangled cable arrangements, which greatly diminish their aesthetic quality. Also in the Chandoline Bridge, seen from certain viewpoints, the double cables in the side spans overlap very unfavorably; single cables would have been much better.

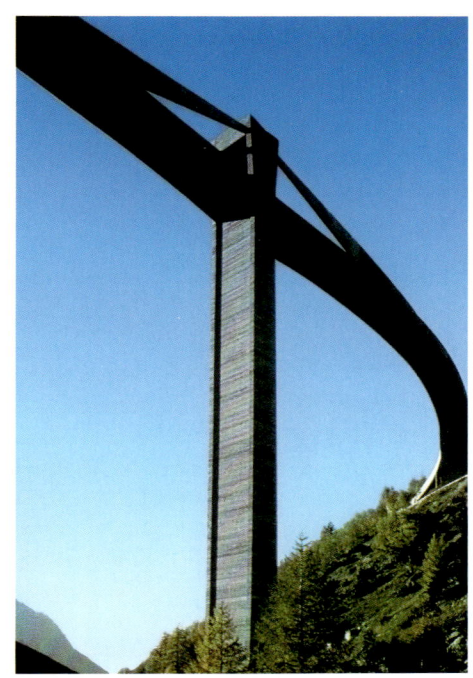

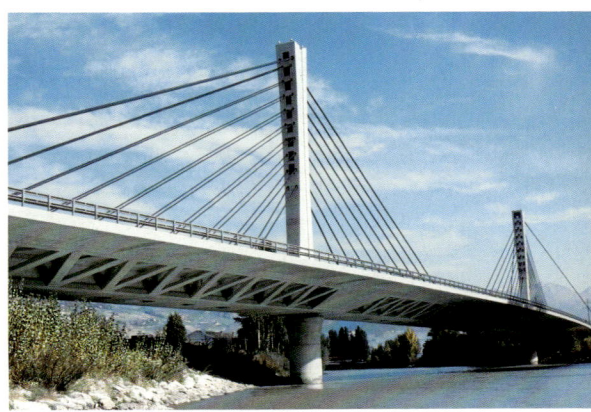

❻ *(top, left)*
Felsenau Bridge, Bern, Switzerland

❼ *(top, right)*
Ganter Bridge, Simplon Pass, Switzerland

❽ *(left)*
Biaschina Bridge, Giornico, Switzerland

❾ *(above)*
Chandoline Bridge, Sion, Switzerland

10 Sunniberg Bridge, Klosters, Switzerland

11 Charles River Bridge, Boston

134 | THE FELIX CANDELA LECTURES

Sunniberg Bridge near Klosters
Built between 1996 and 1998, this bridge is the most prominent visible structure on the extremely costly and busy highway to Davos (fig. 10). Situated in front of Klosters, the famous holiday resort, this bridge occupies a highly prestigious position. The 525-meter-long cable-stayed bridge crosses the Prättigau Valley at a height of about 60 meters in a curve with a radius of 480 meters. I am convinced that this five-span cable-stayed bridge was the most appropriate solution for this setting, conforming to the scale of the landscape and the local topography. It represents the state of the art in bridge design.

The initial concept for the bridge was based exclusively on technical considerations, which led to an extremely slender, joint-free superstructure, Vierendeel-shaped piers, and low pylons. A high technical efficiency was required to achieve the slenderness and transparency of the bridge. The piers, with their naturally rising pylons, and the bridge girder create a consistent, visually clear structural system. All cross sections exhibit the same shallow T-shaped typology. The form of the piers, which corresponds perfectly to the envelope of the moment diagram, visualizes the flow of forces. The rather costly harp arrangement of the cables was necessary to avoid any appearance that the cables were entangled when one drives over the curved bridge. The only ornamentation is the simple but effective recesses in the cross sections of the piers.

Charles River Bridge in Boston
This bridge, which is under construction (2000), posed the most complex functional demands and boundary conditions I have ever encountered: there are eight lanes for Interstate I-93 and two for access ramps; a relatively steep highway gradient; complicated connections (to the tunnel on the southern side and the interstate on the northern side); a skew, pier-free crossing of the river; and a restricted depth of the deck due to the clearance requirements under the bridge. In addition, this bridge plays an exceptionally important role in the urban landscape as a gateway to downtown Boston (fig. 11).

The elaboration of the concept of this bridge reminded me of the inner workings of a clock, where all wheels have to fit together. The conceptual design had to take into consideration several issues. One was the number of pylons. Only one pylon on the northern riverbank would have required a relatively long northern side span, while anchoring the backstay cables in the curved and already split lanes at the beginning of the intersection would have been difficult. Another issue was the different building materials, which included a steel-concrete composite main span and concrete side spans. Owing to the light main span and the heavy side spans, a certain amount of balance could be achieved in the forces of the corresponding cables. This made it possible to avoid a concentration of cables at the ends of the bridge and large bending intervals in the pylons. At the back spans, single cables could be placed with an acceptable spacing between the anchor points.

A third issue was the asymmetry of the bridge cross section. For the half-fan-shaped cable arrangement, only two cable planes are aesthetically satisfactory, and by placing the access ramps outside of the pylons, the span of the floor beams could be reduced considerably. In this way also the gatelike openings of the pylons became narrower and lower. The pylons rise symmetrically above I-93, and the cables can be anchored at the pylon legs and tops, which is important in view of the transverse bending of the pylons. Moreover, the ramps that overlap with the existing bridge can be built after the completion of the new I-93 bridge and after the demolition of the existing one.

The fourth issue was anchoring the cables along the edges of I-93 in the main span. By doing this, the deck girder receives practically no torsion, and its cross section could be designed as a simple and economic T-beam. In the side spans, where the cables are anchored in the median, torsion can be easily resisted by the concrete box girder and the intermediate columns. Anchoring the cables in the median eliminates the possibility of the cables overlapping in the southeastern edge with the existing bridge. The side spans have the same cross section as the connecting elevated highway; the change of cross section takes place at the pylons, where it is not noticed. But the most important advantage of this arrangement of the anchors consists in a reduction of the transverse bending in the pylons caused by the asymmetry of the bridge by more than sixty percent. The fifth issue was the openings in the unusually wide deck, which not only let through some light onto the river, but, above all, prevented the development of unfavorable wind effects.

Nothing in the basic conceptual design of the Charles River Bridge was inspired by purely architectural considerations. The form of the bridge, which offers a fascinating perspective to the innumerable people who use the bridge daily, was based exclusively on functional, structural, and constructional considerations. However, I would like to emphasize that ornamental extras, like the super-elevated tops of the pylons and the structurally unfavorable concave cross sections with the much too small recesses, were not proposed by me. I wanted deeper recesses.

Bridge with a free span of 3,000 meters
Important steps in bridge engineering, such as a significant increase in the free-span length, were always achieved through innovative ideas and not by simply extrapolating existing bridge types. An especially impressive example of a suspension bridge is the George Washington Bridge in New York, completed in 1931, where Ammann doubled the free-span length, a record for his time (fig. 12). This increase of 500 meters has indeed not been surpassed in the history of bridge engineering. According to his famous study on this bridge, published in 1923, Ammann's innovative idea was to reduce the depth (or stiffness) of the bridge girder rather than to increase it for very long-span suspension bridges.

George Washington Bridge with single deck, New York

At present, the Akashi Kaikyo Bridge between Awajishima and Honshu in Kobe city (fig. 13), with a free span of almost 2,000 meters, holds the world record, and a span of 3,000 meters could become reality with the envisaged bridge across the Strait of Messina (fig. 14). At the International Association for Bridge and Structural Engineering / Fédération International de la Précontrainte Symposium at Deauville in 1994, Professor Alberto Castellani from Milan proposed a solution for this bridge consisting of a simple extrapolation of a classical suspension bridge, while Professor Roger Lacroix from Paris suggested a hybrid, but conventional, cable-stayed suspension system. In both cases, the indispensable transverse stiffness, necessary because of the dynamic wind effects, could be achieved only by making the bridge deck much wider than necessary for traffic requirements.

The not-too-convincing extrapolated proposals put forth at the symposium led me to develop one of my own. I also suggested a hybrid system but, for the cable-stayed sections, simply constructable outriggers would be used as pylons. The main advantages of this idea are: during construction the cantilevering cable-stayed sections are stable; sufficient transverse stiffness is attained without expensive widening of the deck; and the construction time can be considerably shortened because the cable-stayed sections can be started before the pylons are completed. In this case, purely technical considerations led to a structural system that exhibits a new and impressive architectural appearance.

⓭ Akashi Kaikyo Bridge, Kobe, Japan

⓮ 3,000-meter span bridge

Especially in the case of difficult and/or long-span bridges, only engineers with sound structural engineering backgrounds can produce good concepts, which exhibit a genuine balance between costs and aesthetics. Although the contributions of architects, designers, and artists are fully desirable in order to refine the raw structural form and to work out certain details, the basic structural concept should be left to qualified engineers. In this sense, the present engineering programs at the universities should be improved. It is certainly not sufficient to provide just scientific knowledge; it is also important to promote and cultivate creative and artistic talents in training structural engineers.

❶
German shell by Dyckerhoff and
Widmann with Anton Tedesko

❷
Mexican shell by Felix Candela

On the Cultural and Social Responsibility of the Structural Engineer

JÖRG SCHLAICH

Felix Candela was, in my estimation, first of all a great, humble, and generous man, and, as an engineer, he strove for beauty through efficiency. Light and efficient structures are becoming more and more timely because they save resources by using fewer materials, thus addressing ecological concerns. They convert material into labor and so fulfill a social function. And since efficient structures can only be achieved by designing them according to their natural flow of forces, they have their own innate beauty and may even make their own cultural statement. So ecological, social, cultural—what could be more timely?

To achieve these structures, Candela focused on concrete shells because double-curved surfaces, more than any other kind of structure, embody efficiency and lightness. They make the best use of the material. However, double-curved sophisticated formwork made of concrete shells is costly, which restricts its use. Too often it loses out to more primitive, bulky, and ugly structures. The challenge for the dedicated engineer is how to overcome this problem.

Candela found his answer with the hypar shell because its double-curved surface can be produced from straight generators. The manufacture of double-curved surfaces, not only for concrete shells but also for grid shells and for membranes with the aim of making them economically feasible, is a subject I would like to address here. In my discussion, I will touch on the interrelation between the geometry, the fabrication, and the load-bearing behavior of different types of double-curved structures, in honor of Felix Candela.

In the mid-1960s, I was working in the office of Fritz Leonhardt, where I had the opportunity to design a very large hypar roof in Hamburg spanning 100 meters. In the course of my research for this work, I was particularly impressed with Candela's ability to translate very complicated situations into simple formulae, as noted in Colin Faber's book on Candela. I was surprised to find that Candela had no problems with unsupported edge beams, while I had difficulty trying to use the same structural form for the Hamburg shell. It took me a long time to understand that because his shell was only about one-third the size of mine, he was able to make it stand up. The importance of scale! The weight of a structure increases to the third power when it is enlarged, the strength only to the second. Candela felt and understood this law and never exaggerated the size.

The early German and Mexican shell builders proudly demonstrated their skill in the same way. Anton Tedesko, who is one of those standing on the German shell (1932; fig. 1), studied with Ulrich Finsterwalder and later brought concrete shells to the U.S., and Felix Candela is certainly one of those standing on the Mexican shell (1953; fig. 2).

I learned from the Hamburg shell that the formwork was more expensive than the shell itself (fig. 3), which is why shells slowly disappeared and were replaced by more primitive types of roofs or by tensile structures, cable nets, and membranes. Unwilling to accept the fact that such beautiful structures would no longer be viable, with the exception of those designed by Heinz Isler, we tried to revive concrete shells by using pneumatic formwork and by prefabrication. The result was a shell built in Stuttgart that is closely modeled after Candela's Xochimilco shell in Mexico (1977; fig. 4). Using a new material—glass-fiber-reinforced concrete—allowed us to reduce the thickness of the shell to 12 millimeters. Thanks to this, we were able to lift large prefabricated shell elements with a simple crane. In looking for a shell design with a repetitive shape as my guide, I found that Candela's Xochimilco shell was the most obvious and beautiful choice.

We prefabricated all eight elements using the same formwork and lifted them to their final position (fig. 5). The only minor difference between the Stuttgart shell and that at Xochimilco was that we wanted to separate the glass fiber part, structurally and visually, from the normal concrete part by introducing, at each support, a small stainless steel ball, a total of eight in all.

❸
Formwork of the Hamburg hypar shell

❹—❺
Stuttgart glass-fiber-reinforced-concrete shell

After we had designed and built this shell, a colleague surprised me by saying, "This is plagiarism. You are stealing the ideas of Felix Candela. You must ask his permission, you must send him an honorarium, and you must apologize, etc. etc.!" It so happened that Candela was teaching in Paris at that time (1977). I telephoned him and said that if he had nothing better to do he might come to Stuttgart because there was something I wanted to show him. So without any formality, he and two students came the next weekend—a six-hour train ride. We went to my home, where I had invited some colleagues to meet him. Somewhat shy about his reaction to the shell, I said: "Now it's time to bring you to the station." But he protested, "No, you promised to show me something." So we went together to my Xochimilco/Stuttgart shell. He laughed, clapped his hands, climbed up and jumped on it, and enjoyed the movement of this shell, because being 12-millimeters thick, it was, of course, very flexible. He seemed very happy about it. With tears in his eyes he said, "It feels very good to know that your own work is useful and interesting enough to bear fruit in the mind and work of some younger colleagues." I was deeply moved, which is why for me Felix Candela was not only a great engineer but also a very humble and generous man.

An alternative to the monolithic concrete shell is the grid shell with triangular mesh of steel slats with an opaque or transparent cladding, preferably of glass. But like the formwork of the concrete shell, there is still the basic problem of creating a double-curved surface on the plane, a concept not unlike the projection of the globe on a plane map. There are multiple solutions but they are all approximations. In the case of the double-curved grid shell with triangular mesh, the sides and nodes of the individual triangles permanently change their lengths and angles, causing a fabrication problem. The best-known effort to minimize these changes was the Geodesic Dome by Buckminster Fuller. An icosahedrom consisting of twenty equal triangles is projected on the inside of a sphere and subdivided into hexagons and twenty pentagons—similar to the pattern of a football—and further subdivided into triangles. Of course, this again is only an approximation, with variable lengths and angles.

Already, as early as 1926, Walter Bauersfeld and Franz Dischinger used this grid as the formwork for a precise spherical concrete shell in their Zeiss Planetarium in Jena, Germany (fig. 6). They thus anticipated Fuller's Geodesic Dome by almost thirty years.

Today's computerized designs and fabrications have made all the efforts of Konrad Wachsmann, Buckminster Fuller, Max Mengeringhausen, Frei Otto, and others obsolete. Recently we designed a glass roof over a courtyard with triangular mesh for optimum load-bearing potential. The project was the DG Bank Building in Berlin at Pariser Platz, next to the Brandenburg Gate (2000). The architect was Frank Gehry. Thanks to computer numerical control (CNC) fabrication, it was possible to have only one axis of symmetry. In addition, each triangle and node permanently varies. This triangular mesh of slats, directly covered with double glazing, allows for an ideal membrane-shell load-bearing behavior, thanks to several stiffening spoked wheels (figs. 7–9).

6
Lattice shell for Zeiss Planetarium, Jena, Germany

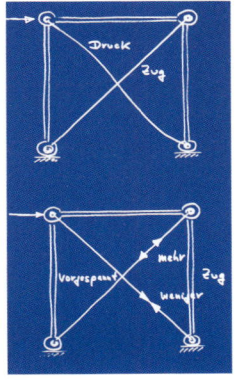

7–9
DG Bank Building, Berlin

10
Drawings of frames

Hopefully, this roof represents a new freedom in design and fabrication, but this freedom must not be misused or it will end in chaos. The former discipline imposed on us by manual fabrication should be replaced by mental self-discipline.

Each solution also has some drawback. If triangular glass panes are used, they are relatively much more expensive than quadrangular ones because they are cut from quadrangles and thus cause substantial waste. Therefore, it makes sense to start with quadrangles and stiffen them with diagonal cables, which, if pretensioned, can efficiently act in tension and compression (fig. 10). A pretensioned tensile element accepts compression by reduction of tension, and thus cables can act in compression without getting slack.

SCHLAICH | 145

⓫ Sieve

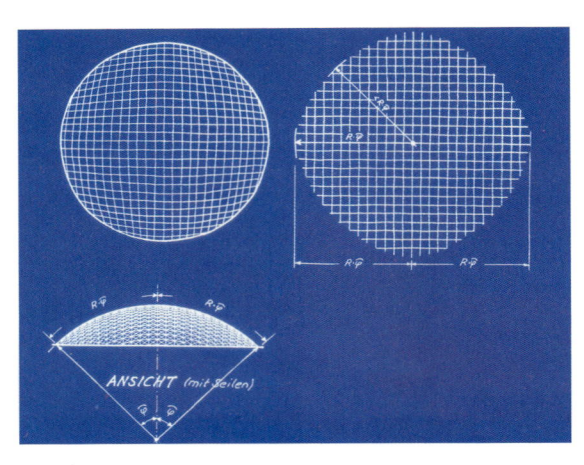

⓬ Drawing of sieves

An ordinary salad sieve reveals a very intelligent dome structure (fig. 11). A simple quadrangular wire mesh with rotational nodes can adopt any double-curved surface by a change of angles but maintaining the equal lengths of all members except those at the edges. Its pattern is determined by adding the number of equal mesh lengths and the two varying end lengths between the last nodes and the edges. Today, this figure can easily be determined by computer. Accordingly, the construction—here exemplified by a dome over an indoor pool at Neckarsulm, Germany—starts with these end members and then continues by adding all the standard slats until the quadrangular grid is completed (1989; fig. 12). Finally, the diagonal cables are introduced. They run continuously from one edge to the other and are pretensioned. The cables consist of two parallel strands so they can be fixed to the nodes, each with one bolt only. The quadrangular glass panes, rhombic in shape and precisely precut, are placed *directly* on the steel grid and fixed only at the edges (figs. 13, 14). To convince the workers on the site of the Neckarsulm dome that these thin cables make sense, we suspended water barrels representing the weight of a local snow load and compared the deformations with and without the diagonal cables (fig. 15).

⓭–⓯
Neckarsulm pool

16–18
Hamburg Historical Museum

Another example where we applied the salad-sieve approach is the roof over the courtyard of the Hamburg Historical Museum. There are two cylindrical shells with different spans joining in a free and smooth transition, the cylindrical parts being additionally stiffened by spoked wheels (or spider nets). In this case, single glazing was sufficient from a thermal point of view, allowing the lightness of this grid shell to become evident (1989; figs. 16–18).

Generally, the four edges of such quadrangular meshes are not in a plane. In the spherical dome with double glazing at Neckarsulm, we used curved glass whereas at Hamburg, which has single glazing, we bent and warped it to fit. Depending on the individual circumstances, these solutions may or may not work. My colleague Hans Schober suggested a simple alternative: the translational surfaces resulting from arbitrary generators and generatrices offer a multitude of shapes that overcome this problem of maintaining equal mesh widths and can be covered with plane glass panes. For a roof at the Berlin Zoo, we joined two cupolas of different sizes, thus, for the first time, successfully demonstrating this approach (1997; figs. 19, 20). Other examples of cylindrical roofs with quadrangular but, by definition, plane meshes include a hanging canopy for the Ulm railway station (1993; fig. 21) and the roof of the Berlin-Spandau station (1998; fig. 22).

A dome or cupola-type structure with a synclastic curvature that acts predominantly by compression must have a monolithic shell surface, a grid with triangular mesh, in order to be stable. On the other hand, a pure tensile structure, say a mechanically (as against pneumatically) pretensioned cable net with an anticlastic curvature, will, under any load, find a state and shape of equilibrium and therefore needs only a quadrangular mesh or two layers of cables. With this in mind we realized that for cable nets with quadrangular mesh we could apply the salad-sieve principle, resulting in an (almost) unlimited variety of shapes.

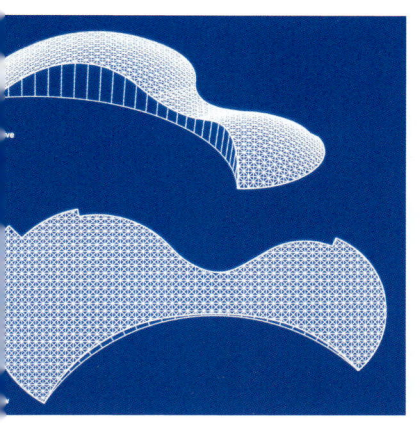

19 Drawing for roof of Berlin Zoo

20 Berlin Zoo

21 Ulm railway station canopy

22 Berlin-Spandau railway station platform roof

SCHLAICH | 149

Our cable-net roof for the Munich Olympics of 1972, designed with Behnisch + Partner and Frei Otto, made full use of this principle, with a unique "roof landscape" (fig. 23). The geometry of such two-layer cable nets with equal distances between the nodes and the turnable joints follows from equilibrium under prestress. Its cutting pattern is defined by the number of equal mesh lengths plus the two varying end lengths between the last nodes and the edge cables, reduced by the elongation due to prestress. An unprecedented accuracy of prefabrication is the result.

For erection, first the primary structure—masts, guy and suspension cables, etc.—is assembled. Then the cable nets with square mesh, prefixed turnable nodes, and edge cables are installed flat while on the ground and then lifted into position. They assume the desired shape by changing angles and by prestress. The roof over an ice-skating rink at Munich, adjacent to the Olympic sports fields (figs. 24, 25), is an example. Whereas the roof for the Olympics was covered with plexiglass, a wooden grid with a textile membrane cover was chosen for the ice-skating rink (fig. 26).

Unfortunately, cable nets with a cover are relatively expensive in comparison to pure membrane structures. The membrane serves two purposes simultaneously: it acts as a cover and it carries the load. Like the cable net, it needs masts and cables. For the fabrication of such textile membranes, which are double-curved but sewn from plane material, we followed the same procedure as the one in making clothes.

㉓ Munich Stadium

24–26
Roof over ice-skating rink, Munich

SCHLAICH | 151

We prefabricated it from strips (fig. 27), using a cutting pattern that was reduced by elongation due to prestress, sewed and welded it together in the shop, packed it there, brought it on site, unfolded it, fixed it to the primary structure, and lifted and prestressed it, as shown in the stadium in Riyadh (1985; figs. 28–29), conceived by Horst Berger. So again and again, like Candela's shells, the fabrication governs the design.

Some of our further membrane structures may show the lightness and natural beauty of these pretensioned textile membrane structures: the Stuttgart stadium (280/200 meters) (1993; fig. 30) with the load-bearing ability of a horizontal spoked wheel, quite similar to the one at Kuala Lumpur; a small but pretty roof over a grandstand at Oldenburg; a roof over an ice-skating rink near Hamburg (1994; fig. 31); and a roof over a pool at Kuala Lumpur (1998; fig. 32), which shows the importance of detail. We always use the seams to follow and visualize the flow of forces to make the interested observer understand and thus like what he or she sees.

Obviously these flexible membranes offer themselves to foldable or convertible structures. For the bullfight arena in Zaragoza, we designed an outer fixed and an inner convertible membrane roof.

27
Cutting pattern for double-curved membrane roof

28–29
Riyadh Stadium

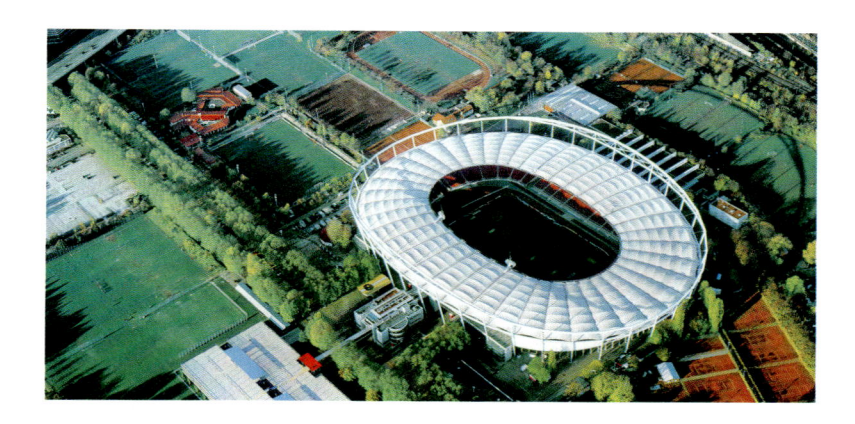

30 Stadium roof, Stuttgart

31 Hamburg ice-skating rink

32 Kuala Lumpur pool

33–36
(this page and opposite)
Zaragoza Stadium

It unfolds like a flower, which may explain why these light structures are sometimes called "natural" (1990; figs. 33–36). For the winter enclosure of the Roman arena in Nîmes, we designed a cushion with synclastic double curvature from pneumatic prestress, which can be unfolded, lifted, fixed to a compression ring, and inflated.

The quality of such membranes today guarantees a lifetime of twenty to twenty-five years if they are carefully designed, manufactured, and maintained. Since we were not sure about that some years ago, we studied an alternative material: thin stainless-steel membranes. Of course, these membranes cannot be folded like textile membranes; they call for another fabricational approach, which we found in plastic deformation of a plane stainless-steel membrane. To test this possibility, we inflated a 0.2 millimeter metal membrane welded from strips. When it stretched, all the wrinkles disappeared, and the cushion assumed an ideal geometrical shape (fig. 37) of optical quality that was more precise than was needed for such a structure. So the idea was to invert the process and produce a concave surface for a solar concentrator with 18 meter diameter (fig. 38). In its focus we installed a receiver for heating helium, which was converted by a Stirling Engine into kinetic energy, thus driving an electrogenerator with 50-kilowatts output at 1 kilowatt hour per square meter of solar insulation. This and the further development of our Dish/Stirling system would be worth its own report. Let me just mention that we continue to improve on it, aiming at indigenous construction for rural application in developing countries.

This leads me to another solar power plant that also goes back to a structural development. In the mid-1970s, we were asked to create cooling towers of a size that could not be made from the usual reinforced concrete. We developed a cable-net cooling tower with a three-layer net and triangular mesh (figs. 39, 40). The size of the triangles changes with height, but, thanks to the rotational shape, not with the perimeter. So there are only two different cables, the two equally

37 Steel cushion

38 Solar mirror

inclined and the meridian cables. These can be simply prefabricated and erected by lifting the net from the ground in its final position and prestressing it.

Cooling towers are vertical tubes, open at their base, which use their internal natural draft to reduce the temperature of the cooling water from the turbine of a coal, oil, or nuclear power plant to ambient, thus emitting about sixty percent of the energy produced. Why not use this draft or upwind to produce electrical energy from the sun? If such a tube is placed at the center of a large, flat, circular glass roof, the sun heats the air underneath the roof and the tube sucks in this warm air so that the upwind can drive a turbine with generator for electricity (fig. 41).

The higher the tube and the larger the roof, the larger the electrical output of such a plant, say 200 megawatts for a 1,000-meter tower and a 5-kilometer roof at 1 KWh/m^2 solar insulation. Our initial design

39–40
Cable net tower

for such high towers began by putting several cable-net cooling towers on top of each other. Later, we found that a "conventional" concrete tube is more economical and durable. To be very efficient, it needed to be stiffened inside by spoked wheels at say five levels. The glass roof governs the overall cost and must, therefore, be simple and robust. Thanks to a fund from our government, we were able to build a small prototype in Spain, which confirmed the feasibility of the concept of the solar chimney or solar updraft tower (figs. 42, 43). The simple glass roof collector can be built by unskilled labor.

The problem of any solar-powered or wind-powered plant is that it can only produce energy during sunny or windy periods. By placing water tubes under the roof, this problem can be overcome. The water stores the heat of the sun and releases it at night, producing twenty-four-hour constant electricity (figs. 44, 45). Detailed cost analysis carried out with the support of utilities have shown that such power

41–43
Solar chimney

plants can easily compete with wind converters, not to mention photovoltaics. They are really sustainable, using primarily glass and cement, which is sand plus energy plus labor.

If we realize that the standard of living of a country is proportional to its energy consumption and that these power plants can be built indigenously in the sunny, developing countries, they offer a unique opportunity. Beyond producing clean and nonpolluting electricity of any amount—eleven percent of the Sahara could satisfy the world's energy consumption— these plants invite the developing countries into the global market as the main energy suppliers. After satisfying their own demand, they could export electricity to the industrialized countries and use this income to buy our goods—to the benefit of this earth and all its inhabitants.

44–45
Diagrams of solar chimneys

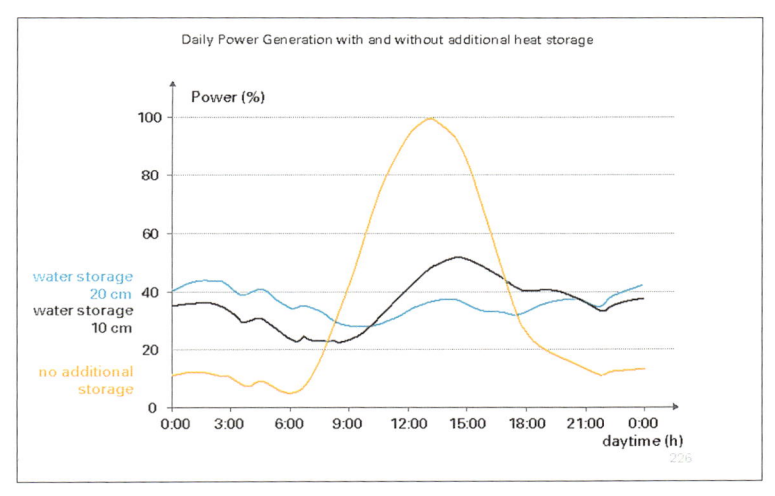

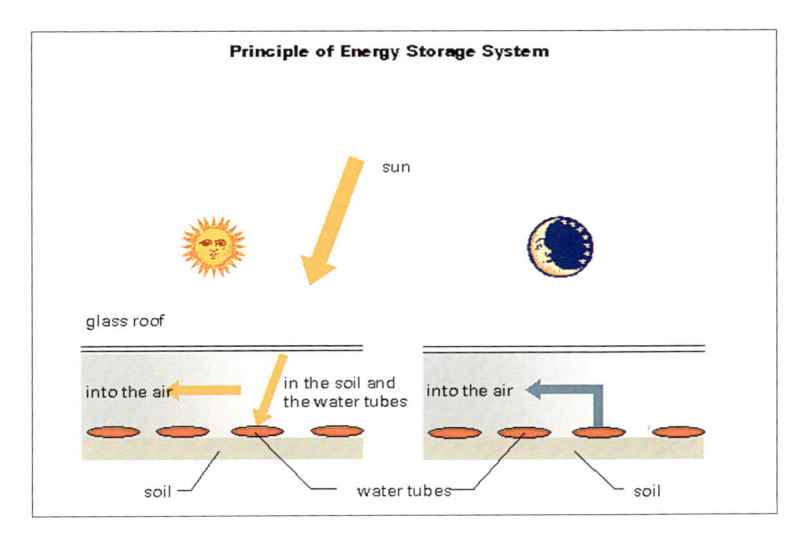

SCHLAICH | 159

❶
Anton Tedesko. Ice-hockey arena, Hershey
Chocolate Company, Hershey, Pa.

Thin-Shell Concrete Structures: The Master Builders[1]

DAVID P. BILLINGTON AND
MARIA M. GARLOCK

Roof structures of thin-shell concrete are a prototypical product of the twentieth century, but they have not been seen much at the start of the present century. This essay focuses on four thin-shell concrete designers, their ideas, works, and relevance for today: Anton Tedesko, Pier Luigi Nervi, Felix Candela, and Heinz Isler. These thin-shell master builders acted as structural engineers, architects, and builders. They had similar education, entrepreneurial characters, and goals. They were all well-educated in structures, and part of their education was critically studying built structures, including their own early works. They acted as individual agents in promoting the work that they proposed to design and see built. Finally, their goal was to design thin shells that would be disciplined by efficiency and economy while expressing an elegance of form that would convey their own unique vision. When a new generation recognizes this tradition, a reassurance in thin-shell concrete shells will be the result.

Anton Tedesko
Rarely can historians attribute to one person the introduction into society of a new and widely useful engineering idea. We have no difficulty, however, in attributing to the structural engineer Anton Tedesko the introduction of thin-shell concrete roof structures into the United States. This achievement merits some reflection not only on the events themselves, but also on the background and personality of the individual engineer.[2]

Three issues central to the history of technology are illustrated by Tedesko's career. The first has to do with the influence of European ideas on the United States; the second with the origins of new engineering ideas; and the third with the tension between techniques and ideas, methods and results, processes and products, and, finally, and more generally, the tensions between means and ends.

Tedesko (1903–1994) studied engineering at the Technological Institute in Vienna, graduating in 1926 with a diploma in civil engineering. He worked in a European practice that strongly emphasized the combination of design and construction within the same company.

For reinforced-concrete structures, this intimate connection led to a practice that stressed competitive designs.

After a year working with a contractor on a large city housing project in Vienna, where he gained field supervision experience, Tedesko traveled to the United States to work for two years. After his return, he became an assistant to Ernst Melan (son of Josef Melan, a prominent nineteenth-century academic engineer), newly appointed professor at the Technological Institute in Vienna. After nine months there, he decided to return full-time to engineering practice, despite Melan's urging that he remain to complete his doctorate. Thus, early in 1930 he joined the designer-builder firm of Dyckerhoff and Widmann in Weisbaden, where his thin-shell practice began.

Dyckerhoff and Widmann had been designing and building reinforced-concrete structures since the late nineteenth century and were continuously engaged in preparing designs to bid competitively against the designs of other designer-builders. When Tedesko began with the firm, it had a remarkable group of structural engineers developing thin-shell concrete roof designs: Franz Dischinger, Ulrich Finsterwalder, Wilhelm Flügge, and Hubert Rüsch. Each of them became world famous for their work in thin shells and concrete structures. The origins of the firm's dedication to thin-concrete shells lay in events of the early 1920s involving domes and barrel vaults.

Based upon their experience with very large domes, i.e., those at Leipzig, and very large barrels, such as the 1928 Market Hall in Frankfurt, Dyckerhoff and Widmann decided to expand their operations abroad. Because of his American experience, Tedesko was sent to the United States in early 1932. There he became affiliated with an older friend from Vienna, John E. Kalinka, who was then working for the Roberts and Schaefer Company in Chicago. This firm of designer-builders was experienced in concrete structures largely for coal-handling facilities and for concrete arch bridges done by the earlier firm of Bush, Roberts, and Schaefer.

The Hayden Planetarium in New York was fittingly the first application on which Tedesko worked. Roberts and Schaefer were invited to serve as consultants for the design and construction of a concrete thin shell to be bid in competition with a Gustavino tile dome alternative. The concrete design, being cheaper, was chosen and constructed by shotcrete, as were the early domes in Germany.

Tedesko's second American thin-shell roof was a barrel shell for the Brook Hall Farm dairy building at the Century of Progress World's Fair, held in 1933 in Chicago. There were five 10.9-meter span, 4.3-meter-wide multi-barrels covering 10.9 x 21.4 meters of interior space. This time Roberts and Schaefer were the engineers, with Tedesko as the principal designer. At the time of demolition, a series of carefully controlled load tests was carried out by the engineers, assisted by the Portland Cement Association and closely observed by engineering staff members at the University of Illinois. These tests characterized the close relationship developed between the association and Tedesko, all with the end result of promoting concrete shells.

During the Depression, most American engineers were uninformed about shell theory and construction practice. The first major thin shell in the United States appeared in 1936, when the Hershey Chocolate Company decided to build an ice-hockey arena (fig. 1). Not only was this 70.7-meter span, 103.7-meter-long roof the largest thin-shell concrete structure on the North American Continent, but it was also "one of several community institutions built by the Hershey interests for the purpose of welding their 2,200 citizen employees into a big happy family." Since the arena holds more than 7,000 people, its seating capacity extended well beyond the chocolate works. After it opened on December 9, 1936, after only eight months of construction, it proved a great success both for the spectators and for the engineering profession. This immense barrel, still in excellent condition after nearly seventy years of service, was designed by Tedesko, who personally supervised the construction. It served as a prototype for a large number of arenas and hangars designed under Tedesko's direction. Its great size, extreme thinness, and impressively short completion time all served to demonstrate that concrete shells could be built in the United States.

Two later thin-shell roofs characterize Tedesko's method of working with well-known architects. In 1953 he was approached by W. C. E. Becker of St. Louis about the design of a groined vault for the new Lambert Field Terminal Building. The architects Hellmuth, Yamasaki & Leinweber had designed a roof of intersecting barrel shells. After studying the roof, Tedesko insisted on adding diagonal ribs above the shell along the valley intersections of the barrels. Although the designer, Minoru Yamasaki, originally planned for a ribless vault, Tedesko's advice was followed, and the structure has stood without any structural problems for more than twenty-five years.

A hyperbolic paraboloidal roof in Denver is the second example of Tedesko's collaboration with a well-known architect on a major thin shell. Here, I. M. Pei wanted a ribless shell of the gable form. Tedesko knew that the flat central part of the shell would be in danger of buckling without a rib so he put in two intersecting heavy, wide, slab bands. These ridge bands were heavily reinforced to inhibit creep. By careful attention to such details, Tedesko ensured that this shell would be free of structural problems.

In summarizing his years of experience with shell roofs, Tedesko stressed "three important points on which depends the success of shells: (1) Designers learn from having experience with full-scale structures; (2) Engineers should visualize the actual construction during the design stage; the economy of shells depends on the close collaboration of designer and builder; (3) Even in working with architects of prestige the engineer must stand firm, making it clear as to what can be done and what should not be done."

Four conclusions can be drawn from this review of Tedesko's work and ideas in thin-shell concrete studies. First, his educational background broadened his vision, emphasized design at least as much as analysis, and included aesthetics as a serious part of structural design. Second, his experience in European practice gave him not only a solid grounding in thin-shell construction, but also a first-hand view of the operation of designer-builders. Third, the history of thin-shell structures shows how new ideas in structural engineering came from designers and especially those with construction experience. Fourth, Tedesko's long career has focused on the design of full-scale works in which behavior, cost, and appearance have been the criteria for success.

Pier Luigi Nervi

Turning from experience in the United States to the Italian Pier Luigi Nervi (1891–1979), we come to an engineer who centered his entire career on aesthetics. There is no doubt whatsoever that Nervi saw himself as an artist whose mission was to create beautiful objects. Beginning to design on his own at the time that the Swiss engineer Robert Maillart's greatest works were appearing, Nervi saw that structure could be art when it arose out of correct form, careful construction practice, and conscious aesthetic intention. During the 1930s, Nervi was designing and building large concrete structures, mostly as a result of winning cost competitions, although his intellectual bent was for reflection and aesthetics. Whereas Dischinger and Finsterwalder were writing about domes and barrels designed on the basis of the mathematics theories they had developed, Nervi was writing such articles as "The Art and Technique of Building," "Thoughts on Engineering," "Problems of Architectural Achievement," and "Technology and the New Aesthetic Direction." He wrote no treatises on scientific analyses.[3]

Nervi began practicing after graduation from the University of Bologna in civil engineering in 1913, but it was not until 1932, at the age of forty-one, that he completed a major structure on his own. During those nearly twenty years of practice, however, he gained design and field experience with reinforced concrete, so that when he first began to design and build his own works they were the product of considerable maturity. The most spectacular of these are certainly the domes and barrel shells he built between 1935 and 1959. These are also the works that best illustrate Nervi's preoccupation with very simple overall shapes made up of an interplay of individual elements. In his domes, these elements are ribs, which make the overall dome both stable and light. This is precisely the tradition that one sees from inside of certain Roman and Italian Renaissance domes.

With this historical background, Nervi approached reinforced concrete in just the way Maillart had: both as a builder of competitive structures and as a designer of new forms. As Nervi put it, his early experiences "had formed in me a habit of searching for solutions that were intrinsically and constructionally the most economic, a habit which the many succeeding competition tenders (almost the totality of my projects) have only succeeded in strengthening." Nervi's whole outlook was, therefore, influenced by the search for economy. He would have had almost no chance to build had his designs not been the cheapest. At the same time, this economy was, for Nervi, intimately connected with finding "the method of bringing dead and live loads down to the foundations... with the minimum use of materials." Economy of cost and efficiency of materials were, however, never enough, for as he continued, "I still remember the long and patient work to find an agreement between the static necessities... and the desire to obtain something which for me would have a satisfying appearance." Nervi continually emphasized that although in structural design the study of external loads and internal resisting forces (the statics) "offers a definite direction,... the detailing of forms and their interrelationship is a personal choice."

Nervi also realized clearly that his own vision had been formed in his native Italy and could only with great difficulty be transferred to another country. "Many times," he wrote, "I have refused to accept commissions... in countries with whose possibilities for building [large structures] I was not familiar in order to avoid running the risk of designing shapes and structures which might prove impossible to build." Once world famous, Nervi did accept a few prestigious commissions abroad, the best known being the UNESCO center in Paris in 1957 and the George Washington Bridge Bus Station in New York in 1963. However, these do not reflect Nervi's greatness as a structural artist, even though the UNESCO buildings are examples of collaboration with architects. It is to his Italian works that we must look for an understanding of Nervi's art.

In the final and greatest masterpiece of his diagonally ribbed style, Nervi's 1957 design for the Little Sports Palace in Rome (fig. 2), the double-rib system creates a decorative pattern out of a technically superior design idea. That idea is related to the major technical problem in thin-shell domes—buckling. It is possible to construct immense ribless domes of exceptional thinness in which under gravity the internal compression stresses are small. But even small stresses can cause deformations, which in a very thin surface can change the geometry enough to buckle the shell and result in collapse.

By 1948 Nervi had found this method of enhancing the safety of domes without increasing mass; his system in Italy was economical, and it liberated his imagination to express a variety of spectacular forms. With the Little Sports Palace, Nervi achieved a high point in the structural art of building in concrete, and he achieved it on his own by learning from his past works and by striving always for beauty and economy.

In many of Nervi's buildings there is a collaborating architect, but the constant development in style is entirely Nervi's. He frequently worked with sensitive architects, but the structural design for his Italian works was always his own. Nervi's works are greatest when the architectural requirements converge most closely with pure structural

❷
Pier Luigi Nervi. Little Sports Palace, Rome

ones. Where there is a complexity of functions, Nervi's art is compromised. For example, the Little Sports Palace is almost pure structure, and it shows that purity both from within and from without. By contrast, the Large Sports Palace, which Nervi designed for the 1960 Olympics, is so large that its auxiliary functions produced a surrounding building complex that destroys the exterior structural expression, and partly disrupts the visual logic of the interior structure as well. The main domed space is spectacular, but the approach view is more like that of some giant water tank than of a buttressed Nervi dome.

By 1948 Nervi had already fulfilled the architectural historian Sigfried Giedion's prophecy that thin shell roofs would be the "solution of the vaulting problem for our period." What is essential to see now is that Nervi's direction was not the only one. As he himself continually stressed, structural art stimulated many possible solutions to the same technical problem. To see this diversity of solutions we shall turn next to other designers who were at the same time studying other ways to achieve technical excellence, competitive economy, and visual richness. Along with Nervi and Tedesko, it was the designers from Spain who led the way before World War II and immediately afterward.

Felix Candela

Modern vault designers in Spain expressed visually the structural idea of thinness by emphasizing smooth, ribless surfaces, and by searching more widely for forms never used before in large buildings. The three designers who best characterize this modern "Spanish School" are the Spanish-born Antonio Gaudí (1852–1926), Eduardo Torroja (1899–1961), and Felix Candela (1910–1997), who was exiled to Mexico in 1939. After World War II, American designers became aware of the works of these Spanish designers. Most immediately influential was Felix Candela, whose hyperbolic paraboloid roofs showed radically new possibilities for appealing shapes that were at the same time both structurally sound and constructionally economical. A true master builder, Candela was a builder, an architect, and a structural engineer. His wide variety of completed structures in Mexico stimulated designers in the United States. The social, political, and economical context in Mexico during Candela's time as a designer set the stage for his success.[4]

Candela was already forty years old before his career as a designer and builder began, and yet by then he was so well prepared that after only four years he was clearly recognized as a master. This rise owes itself to a series of events as well as to his personality. One of these events has to do with the cultural renaissance in Spain that came to be associated with the 1898 loss of the war with the United States. The major figure associated with the flowering of intellectual life in the following half-century was the Spanish philosopher José Ortega y Gasset (1883–1955), whose writing strongly influenced the generation that came to maturity in the 1930s. Candela's first major writing on structural design, "Towards a New Philosophy of Structures," contains a series of quotations from Ortega, but even more importantly his ideas on design seem to be closely similar to those expressed in 1925 by Ortega in his famous essay, "The Dehumanization of Art," in the sense that Candela departed from the traditional way of thinking about structures.

Both Ortega and Candela present their ideas as reactions to dominant themes in nineteenth-century European thinking. Spain had played a small role in the intellectual ferment in Europe over the two centuries in which the Industrial Revolution transformed Western culture. Thus when Ortega began to write, it was as if from another world that his voice resounded. In his analysis of modern art, Ortega thought mainly of music, painting, poetry, and the theater, but his insight is so profound and prophetic that it applied directly to structural art as well.

A modern art had arisen in Spain before the Civil War, and it was during this time that the structural artists in Spain began to flourish. If Ortega was perhaps the major spokesman for this Spanish Renaissance, then artists such as Pablo Picasso, Joan Miró, Salvador Dalí, and Pablo Casals provided evidence for its vigor and originality. For Candela, the two major figures were Antonio Gaudí and Eduardo Torroja. Like the painters and musicians, Torroja has since become recognized as a world leader in twentieth-century structural art because he created new

❸
Felix Candela. Los Manantiales restaurant, Xochimilco, Mexico

❹
Felix Candela. Open chapel, Lomas de Cuernavaca, Mexico

forms that have surprised and enriched engineers, architects, and the general public. It is not that Candela was directly influenced by Torroja or by Gaudí, but rather the fact that all three designers created such similar things shows the strength of a new Spanish tradition in the same way as do the collective paintings of Picasso, Miró, and Dalí.

Candela immigrated to Mexico after the Spanish Civil War. For him, the most important features of Mexico, especially after World War II, were its openness to new ideas and its accelerated modernization. Between the years 1950 and 1970, the years of Candela's Mexican building career, the country went through an industrial revolution. For example, its wheat production doubled, and there was almost no inflation. Within this economic context, Mexico, and especially Mexico City, underwent an unprecedented building boom that provided Candela with a unique opportunity to try out new ideas and to see them built rapidly and in large numbers. In Mexico of the 1950s there were no restrictive codes that made thin shells hard to build; labor was relatively inexpensive so that new construction methods did not unduly burden costs. Candela's hyperbolic paraboloid shells show the variety of shapes possible with one geometric form when in the hands of a strong entrepreneurial structural engineer who clearly understood construction and was sensitive to aesthetics. Examples of his works are the restaurant in Xochimilco (1958; fig. 3) and the chapel at Lomas de Cuernavaca (1958–59; fig. 4).

Heinz Isler

In 1958 Eduardo Torroja organized and oversaw the first congress of a new organization, the International Association for Shell Structures (IASS). Torroja's goal for the congress was to present new developments in thin shells rather than to summarize the major works completed since the 1920s.[5] In 1948 Torroja had stimulated Pierre Lardy (1903–1958), professor of concrete structures at the Federal Technological Institute in Zurich, to develop and support a models laboratory in Zurich. Now he was encouraging structural engineers to exploit more fully the potential for thin-shell concrete structures. In retrospect, one event from this congress stands out as surprising and profound: the late arrival of one of Lardy's former students, Heinz Isler (born 1926). He had come directly from military service and presented the last paper of the congress. A total of twenty-five papers were presented, all of which later appeared in the first eight issues of the association's new *Bulletin*.

The first twenty-four papers were not unlike most conference papers—presentations of individual structures or groups of them, which elicited a few questions from the audience. Nothing quite prepared the reader for the twenty-fifth paper, by Isler. It was entitled simply, "New Shapes for Shells," and it contained slightly more than one page of text and nine illustrations. In the text, Isler briefly described several ways to arrive at shell shapes, which he showed by illustrating some models and a few completed concrete shells. A full page showed thirty-nine possible shapes with a fortieth replaced by the abbreviation "etc." The implication is of unlimited possibilities.

The paper had an immediate impact and brought forth a rash of discussion, which in print was about five times as long as the actual text. The quality of the discussers was as remarkable as the relative length of the commentary, which was dominated by the most distinguished designers present: Torroja himself, Nicholas Esquillian of France, and Ove Arup of Great Britain. Isler's brief presentation struck three themes that the discussers addressed: that new shapes come from simple models; that the resulting nongeometric shapes could be economically built; and that the shapes were designed without reference to architecture.

Isler noted that technical education in mathematics and geometrics did not directly provide the means for studying new shapes. What is important he stressed, are *physical analogies,* such as the inflated membrane or the draped cloth. But that was only a first part of the problem; a second part was the cost. Isler said: "We use a system which may have a few new ideas." These consisted of very carefully made glue-laminated wood arch pieces placed on light, tubular scaffolding and supporting wooden boards, over which fiberboard insulation rests as formwork. In short, Isler had used both new shapes and new construction methods at the same time.

Heinz Isler had studied geometric forms and, finding them unsatisfactory, arrived at two revolutionary ideas, both of which have an aesthetic origin that connects them to Lardy's influence. But both are fully the product of Isler's own imagination. At twenty-eight, the same age at which Maillart conceived the hollow box in concrete, Isler pioneered his first idea, the pneumatic form. He devised an experiment by building a simple wood frame that held a rubber membrane, which he inflated to achieve a pillow form (see page 90). This device gave him a shape that was in pure tension under pressure, loading from below; thus, such a loading, from above, would put that same shape into pure compression. That load, on a flat shell, is so close to gravity loading (purely vertical) that Isler was sure the form would still be in pure compression and thus be ideal for reinforced concrete. By 1955 he successfully completed a pneumatic structure for the Trösch Company, which led to other commissions for industrial buildings.

Isler's second great discovery occurred in the early summer of 1955. On a building site he saw wet burlap draped over a mesh of steel bars in such a way that within one square opening the cloth hung in a domelike shape solely under its own weight (see page 93). From this observation, he immediately concluded that the cloth carried itself in pure tension so that when reversed, it could be the form for a concrete shell under pure compression. But to utilize this concept for actual construction required more time, even though Isler was able to describe it at the IASS Madrid conference of 1959.

One other key event during the 1950s was Isler's collaboration with the nearby Langenthal building firm of Bösiger, whose personnel he trained in his new building methods. Unlike Nervi and Candela, Isler never had a construction business, so he had to design his shells and the procedure for building them himself, and then, like Tedesko, convince a contractor to follow his ideas carefully. This led to both high-quality construction and competitive-cost structures.

Another incident helped Isler achieve his mature structures of the late 1960s. In the early 1960s, he saw in a Zurich store a book cover with a photograph of the Xochimilco restaurant of Felix Candela; the forms that appeared exhibited a completely new shape—almost paper thin—and caused Isler to think about the expression of such thinness.

Isler's unique professional constructions, the most prototypical of his mature designs, are the hanging-membrane reversed shells at Heimberg for covering tennis courts (1979; fig. 5). Isler identified the tennis-court form, on a ground plan of 48 x 18.6 meters, as the most technically difficult one for the hanging-membrane reversed shell. This is because the great length to width ratio (about 8/3) makes the shell act more like a single, wide, flat, and very thin arch than a series of

❺
Heinz Isler. Heimberg Tennis Center, Heimberg, Switzerland

intersecting arches acting in different directions, as in his square-plan swimming-pool shells. Isler conquered that difficulty by his highly disciplined shaping that produced a form of great stiffness. His long period of study for this new form predicted such rigidity, and his full-scale measurements confirmed it. This result verified the strict technical discipline of efficiency for the tennis-court roofs, whereas the discipline of economy was assured by the construction procedure he developed in the 1950s.

"Symbolic significance" means those aspects of a structure that make it a work of art within the disciplines of efficiency (minimum materials) and economy (competitive costs). Within these disciplines, Isler seeks to make an elegant form, one that carries the personal expression of the artist, which at Heimberg we can approach through an analysis that gives meaning to the design beyond its efficiency and economy.

Johan Huizinga, a Dutch humanist, proposed a theory that "the profound affinity between play and order is perhaps the reason why play...seems to be to such a large extent in the field of aesthetics." This gives a good insight into the playing of tennis and the making of form: "Play has a tendency to be beautiful." Here again Huizinga is talking about creations and performances by people, not by nature. These are conscious activities. His strongest insight comes with the supposition that "it may be that this aesthetic factor [play] is identical with the impulse to create orderly form, which animates play in all its aspects." Isler credits Lardy's teaching to this impulse. The same impulse is symbolized almost perfectly in the out-of-the-way thin-shell concrete roofs where, as Huizinga would say, "the least deviation from it [in order] 'spoils the game,'" the precision essential for the game of tennis and the shapes of shells.

Conclusion
Structural art, having grown up with the Industrial Revolution, is symbolic of free democratic societies. Its goals of efficiency, economy, and elegance correspond to those of societies based upon conservation of limited resources, accountability of public funds, and responsibility for encouraging the creation and preservation of art. The Swiss shells of Isler, such as those at Heimberg, exemplify these goals. In addition, they illustrate the influence of a good teacher, Pierre Lardy, who instilled in his students the universal goal of "encouraging us to find and apply aesthetics from within us." This goal will always meet resistance from those who are not concerned about minimizing the use of natural resources, being accountable for construction funds, and finding a new expression in art. For those who are concerned, Isler's shells point to a new future.

Notes

1. A similar version of this text was published in the *Journal of the International Association for Shell and Spatial Structures,* vol. 45, no. 3, in December 2004, pp. 147–55, and is reprinted here with permission.
2. D. P. Billington, "Anton Tedesko: Thin Shells and Esthetics," *Journal of the Structural Division, American Society of Civil Engineers* 108, no. ST11 (November 1982): 2539–54; and "Hershey Arena: Anton Tedesko's Pioneering Form," *Journal of Structural Engineering* 129, no. 3 (March 2003): 278–85 (with Edmond P. Saliklis).
3. Pier Luigi Nervi, *Aesthetics and Technology in Building* (Cambridge, Mass.: Harvard University Press, 1965) and David P. Billington, *The Tower and the Bridge: The New Art of Structural Engineering* (New York: Basic Books, 1983), 176–83.
4. Colin Faber, *Candela: The Shell Builder* (New York: Reinhold, 1963), and David P. Billington, "Felix Candela and Structural Art," *Bulletin of the International Association for Shell Structures* (January 1986): 5–10.
5. John Chilton, *Heinz Isler* (London: Thomas Telford, 2000), and David P. Billington, *The Art of Structural Design: A Swiss Legacy* (Princeton: Princeton University Art Museum, 2003), 128–62.

Authors' Biographies

Stanford Anderson was born in 1934 in Redwood Falls, Minnesota. He studied architecture and history at the University of Minnesota (BA), the University of California at Berkeley (M Arch), and Columbia University (PhD). He is professor of history and architecture at the Massachusetts Institute of Technology and is former head of the department of architecture. His books include *On Streets* (1986), *Style-Architecture and Building-Art* (1993), *Peter Behrens and a New Architecture for the Twentieth Century* (2000), *and Eladio Dieste: Innovation in Structural Art* (2004). He lives in Boston, Massachusetts.

Cecil Balmond was born in 1943 in Sri Lanka. He studied civil engineering at the University of Southampton (BSc) and at Imperial College of Science, London (MSc). He has worked at Ove Arup & Partners since 1968 and is now deputy chairman of the firm. Balmond has collaborated with architects Rem Koolhaas, Daniel Libeskind, and Toyo Ito on numerous building projects; he has also led the design of the Battersea Powerstation redevelopment in London and the Coimbra Footbridge, Coimbra, Portugal. In 1998 he published *Number 9* and in 2002 a collection of his work in the book *Informal*, which won the 2005 Banister Fletcher Prize for the best book on architecture. He is Paul Philippe Cret Practice Professor of Architecture at PennDesign, Philadelphia, and lives in London, England.

David P. Billington was born in 1927 in Bryn Mawr, Pennsylvania. He studied civil engineering in Belgium on a Fulbright Fellowship and at Princeton University. He is Gordon Y.S. Wu Professor of Engineering in the Department of Civil and Environmental Engineering at Princeton University and serves as the director of the Program in Architecture and Engineering. He has published numerous books, including *The Tower and the Bridge* (1983), *Robert Maillart and the Art of Reinforced Concrete* (1990), and *The Art of Structural Design: A Swiss Legacy* (2003). He lives in Princeton, New Jersey.

Maria M. Garlock was born in 1969 in La Plata, Argentina. She studied civil engineering at Lehigh University, Bethlehem, Pa. (BS), and at Cornell University (MS) and received her PhD in structural engineering from Lehigh University in 2003. She has been an assistant professor in the Department of Civil and Environmental Engineering at Princeton University since 2003. She has contributed to numerous journal and conference publications. She lives in Princeton, New Jersey.

Heinz Isler was born in 1926 near Zurich, Switzerland. He studied civil engineering at the Swiss Federal Institute of Technology (ETH), Zurich, under Professor Pierre Lardy, graduating in 1950. He opened his independent office in Burgdorf, Switzerland, in 1954. His career began with an investigation into thin concrete shells and has led to the invention of methods for creating forms. In 1972, working with Jürgen Joedicke and Günter Behnisch, he submitted the prize-winning design for the main pavilion of the Munich Olympics. He lives with his wife, who has been instrumental in his career, in Zuzwil, Switzerland.

Mamoru Kawaguchi was born in 1932 in Fukui, Japan. He studied at Fukui University and received his M.Eng. and Dr.Eng. degrees from the University of Tokyo. He is Professor Emeritus at Hosei University and has collaborated on major structures with Kenzo Tange, Arata Isozaki, and other leading architects. He is the inventor of the Pantadome system, which has been applied to many major spatial structures. His publications include *Mechanism of Building Structures* (1991) and *Structural Perspectives* (1994). He lives in Tokyo, Japan.

Christian Menn was born in 1927 in Meiringen, Switzerland. He studied civil engineering at the Swiss Federal Institute of Technology (ETH), Zurich, and completed his doctoral degree under the instruction of Professor Pierre Lardy. He soon began designing long-span concrete bridges and utilized the new material—prestressed concrete—to introduce innovative shapes and construction techniques to the field of bridge design. He formed his independent consulting practice in 1957, and in 1971 he became Professor of Structural Engineering at the ETH, from which he retired in 1992. Since his retirement he has been working as an individual consultant. He lives in Chur, Switzerland.

Guy J. P. Nordenson was born in 1955 in Neuilly sur Seine, France. He studied literature, philosophy, and civil engineering at MIT and structural engineering at the University of California at Berkeley and was a Loeb Fellow at Harvard University. He worked in San Francisco for Forell/Elsesser and in New York for Weidlinger Associates before establishing the New York office of Ove Arup & Partners in 1987. In 1997 he founded his own firm, Guy Nordenson and Associates. He is a professor of architecture and structural engineering at Princeton University. In 2003 he published *Tall Buildings* and the following year he co-curated, with Terence Riley, the MoMA exhibition of the same title; in 2004 he published *WTC Emergency Building Damage Assessments*. He lives in New York City.

Leslie E. Robertson was born in 1928 in Los Angeles, California. He studied engineering at the University of California, Berkeley, and began his career in San Francisco working with John A. Blume & Associates. In 1958 he moved to Seattle, Washington, where he worked with Worthington, Skilling, Helle and Jackson, and in 1964, as a Partner, he moved to New York City to lead the structural design of the World Trade Center complex with the architect Minoru Yamasaki. In 1986 in New York, he formed his own practice, Leslie E. Robertson Associates. Robertson has collaborated with architects Max Abramovitz, Gunnar Birkerts, Henry N. Cobb, Romaldo Giurgola, Philip Johnson, Kiyonori Kikutake, William Pedersen, I. M. Pei, Robert Venturi, and many others on major tall buildings and other structures around the world. He teaches at Princeton University and lives in New York City.

Jörg Schlaich was born in 1934 in Stetten im Remstal, Germany. He studied architecture and civil engineering at the University of Stuttgart, civil engineering at the University of Berlin, and received a MSc from Case Institute of Technology, Cleveland, and a doctorate from the University of Stuttgart in 1963. As a partner of Leonhardt und Andrä, he was responsible for the design of the Olympic Stadium, Munich (1972). From 1974 to 2000 he was a full professor at the University of Stuttgart and director of the Institute for Structural Design. He founded his own firm, Schlaich, Bergermann und Partner, in 1980, out of which have come numerous innovations in structural design. In 2004 the German Architectural Museum (DAM), Frankfurt, published *Leicht Weit / Light Structures* by Jörg Schlaich and Rudolf Bergermann. He lives in Stuttgart, Germany.

Acknowledgments

This book testifies to the generosity and cultural commitment of the structural engineers whose voices and works occupy these pages. The role of the structural engineer in modern societies is cast somewhere between that of the scientist and the artist, the accompanist and the soloist. How the seven engineers presented in this book have negotiated the balance of roles varies, but at heart all have committed their lives to their work, and for this I hope they will be better acknowledged thanks to this publication.

The Candela Lecture series was initiated by me in late 1996 as a cooperative effort of the Structural Engineers Association of New York (SEAoNY), The Museum of Modern Art's Department of Architecture and Design, and the School of Architecture at Princeton University and at MIT. For SEAoNY, the lectures were a means of providing an occasion to hear great engineers speak to a wide and cultured audience. For MoMA, Princeton, and MIT it was an excellent means of establishing a continuing discussion about the place of talented and creative structural engineers in the arts and architecture. Felix Candela was himself pleased when I suggested the idea to him in early 1997 and hoped to give the first lecture. When he fell ill later that year, he asked that David Billington speak on his behalf, which he did with great eloquence.

Over the years the presidents and directors of SEAoNY supported the lecture series and offered important suggestions for the choice of speakers. These officers included Rick Mahoney, Ramon Gilsanz, Vicky Arbitrio, and Christa Basel. From MIT, Professors Herbert Einstein, Jerome Connor, Stanford Anderson, and Anne Simunovic; from Princeton, in Architecture, deans Ralph Lerner and Stan Allen and Cynthia Nelson, Linda Greiner, and Fran Corcione, as well as the chairs of Civil Engineering Peter Jaffe and Michael Celia and Professor Maria Garlock, all added their support and ideas to the planning. At MoMA Curbie Oestreich, Rachel Judlowe, Laura Beiles, and Linda Roby helped organize the events each year. In later years the addition to the nominating committee of Antoine Picon from Harvard and Paris, Ricky Burdett from the London School of Economics, and Mutsuro Sasaki from Nagoya University and Tokyo helped sharpen the focus and choice of speakers.

The transcripts of each taped lecture were prepared by Amy Wickner, which were then partially edited by the authors but mostly fashioned to their clear, current form by the editing skill of Joanne Greenspun and David Frankel of MoMA's Department of Publications. Designer Tony Drobinski oversaw the production with rigor and elegance, and Elisa Frolich kept us more or less on schedule and organized. I want to thank Michael Maegraith, who initially agreed to go forward with this book, and Christopher Hudson, who, as his successor in the Department of Publications, supported it to its completion. Katherine Riley in my office organized the photographs and fine-tuned the drafts and layouts. Without her help the book would not have come together as well as it did.

Finally, I would like to thank Terence Riley, former Chief Curator of MoMA's Department of Architecture and Design, for his moral and intellectual support in making both the lecture series and this book possible. He fully understood not only the beauty and rigor of the work of Felix Candela but also the cultural import of that work as inspiration and reference for the succeeding generation of creative structural engineers and those that will follow.

Guy Nordenson

Photograph Credits

Photographs of works reproduced in this volume have been provided in most cases by the authors of the texts. Copyright and/or other photograph credits, listed below, appear at the request of the owners of individual works.

Vicente del Amo, Seville: pp. 35 below and left, 36

© 2007 Artists Rights Society (ARS), New York/ADAGP, Paris: p. 19

© 2007 Artists Rights Society (ARS), New York /ADAGP, Paris / Estate of Marcel Duchamp: p. 25

Arup, London: p. 18

Musée Bartholdi, Colmar (reproduction Christian Kempf): p. 24

David P. Billington: pp. 13, 17

Dorothy Candela: frontispiece, pp. 6, 8, 10, 11, 86

Dieste y Montañez, Montevideo: pp. 33 left, 34 above and below, 35 right and above, 38, 40 left, 41 right, 42 left, 43 top left and center right, p. 46

Wolfgang Hoyt Esto: p. 21 right

Historic American Engineering Record; photograph by Jet Lowe, February 1984: p. 22

Heinz Isler: p. 17

Guy Nordenson and Associates: p. 20

Shinkenchika-sha, Tokyo: pp. 30, 33 right, 34 left, 37, 40 below left and below, p. 41 right and below, 42 right, 43 center left and bottom left, 44, 45

Mary Ann Sullivan, Bluffton University, Bluffton, Ohio: p. 12 right

From Dorothy Harley Eber, *Genius at Work: Images of Alexander Graham Bell* (New York: The Viking Press, 1982): p. 66

From José Antonio Fernández Ordóñez and José Ramón Navarro Vera, *Eduardo Torroja* (Madrid: Ediciones Pronaos, 1999): p. 12 left

From *Tall Buildings* (New York: The Museum of Modern Art, 2003): p. 21 left

Index

Numbers in italics indicate pages on which illustrations occur.

A

Alberti, Leon Battista, 124
Allen, Stan, 15
Ammann, Othmar H., 12, 13, 128; Bronx Whitestone Bridge, New York, 13; George Washington Bridge, New York, 13, 136, *137*; Verrazano Narrows Bridge, New York, 13
Ammann, Robert, 59, 60
Anderson, Stanford, 15, 16
Arup, Ove, 9, 14, 15, 50, 51, 170; Penguin Pool, Regent's Park, London, 50

B

Balmond, Cecil, 15, 21, 22, 25, 49–65; Braga Stadium, 22; Carlsberg Brewery, Northampton, UK, 50, 50–51; Casa da Musica, Porto, Portugal, 55, *55*; CCTV Headquarters, Beijing, 22, 55–56, *56*; Dutch Telecom temporary exhibition stand, *52*, 53; Grand Palais, Lille, 54; House, Bordeaux, 22, 53, *53*; Kunsthal, Rotterdam, 54; Neue Staatsgalerie, Stuttgart, *51*, 51–52; No. 1 Poultry, City of London, 53; Pavilion for Serpentine Gallery, London, 57–59, *58*; Polytechnic, Singapore, 53; Portuguese Pavilion, Hannover, Germany, 62; Power station, Battersea, London, 64–65, *65*; Science Library, Irving, California, 53; Seattle Public Library, 54, 55; Victoria and Albert Museum extension, London, 22, 59, 60, *60*, 61, *61*
Balz, Michael: Open-air theater, Grotzingen, Germany, 93–94, *94*, 95
Bartholdi, Auguste: Statue of Liberty, *23*
Bauersfeld, Walter: Zeiss Planetarium, Jena, Germany, 144, *144*
Becker, W. C. E., 163
Behnisch, Gunter, 17; Stadium roof, Munich Olympics, 17, 150, *151*
Bell, Alexander Graham, 66, 67
Berger, Horst, Stadium, Riyadh, 152, *152*
Bergermann, Rudolf, 17
Berlin, Isaiah, 14
Billington, David P., 12, 13, 15
Birkerts, Gunnar, 19, 82, 83
Bösiger firm, Langenthal, 171
Breuer, Marcel, 19
Brunelleschi, Filippo, 124
Bunshaft, Gordon, 19; Beinecke Rare Book Library, New Haven, 21, *21*
Burdett, Ricky, 15
Burgee, John, 76, 83; AT&T Headquarters, New York, 76, *76*
Bush, Roberts, and Schaefer, 162

C

Candela, Dorothy, 10
Candela, Felix, 9–16, 19, 22, 53, 87, 103, 141–43, 152, 161, 167–69, 171; Bacardi Rum Factory, Matamoros, 9; Chapel, Lomas de Cuernavaca, Mexico, 14, 87, *168*, 169; Church of La Virgen Milagrosa, Colonia Vertiz Navarte, Mexico City, 10, *10*; Church of San

José Obrero, Monterrey, Mexico, 86, 87; Cosmic Rays Pavilion, National University of Mexico, Mexico City, 8, 9; Dance Hall, Acapulco, 87; Fernandez Factory, 9; Lake Tequesquitengo, Morelos, signpost, 8, 9; Los Manantiales restaurant, Xochimilco, Mexico, 18, *168*, 169, 171, *187*; Mexican shell, Xochimilco, *140*, *142*; Sales Office, Guadalajara, 10, *11*, 14; Sports Palace, Palacio de los Deportes, 10, *10*, 11, *11*
Casals, Pablo, 167
Castañeda, Enrique, 10
Castellani, Alberto, 137
Chan, Edwin, 18
Curtis and Davis, 68; IBM Building, Pittsburgh, *67*, 68, *68*

D

Dalí, Salvador, 167, 169
Davenport, Alan G., 70
Dieste y Montañez, 45
Dieste, Antonio, 45
Dieste, Edouardo, 45
Dieste, Eladio, 15, 16, 25, *31*, 31–47; Barbieri e Leggire Service Station, Salto, 35, *35*; CADYL horizontal silo, Young, Uruguay, 45, *46*; Church of Christ the Worker, Atlántida, 41–42, *41–42*, 44; Church of San Pedro, Durazno, 42, *43*, 44, *46*; Dieste's house, Montevideo, *40*, 40–41; Massaro Agroindustries, Joanico, Uruguay, *30*, 31–34, *33*, *34*; Metro maintenance hanger, Rio de Janeiro, 35, *35*; Municipal bus terminal, Salto, 34–35, *35*; Navios silos, Nueva Palmira, 44, *45*; Port Warehouse, Montevideo, 36–38, *37–38*; Television tower, Maldonado, 44; TEM factory, Montevideo, *36*, 36–37; Water tower, Las Vegas, Uruguay, 44, *44*
Dieste, Rafael, 44
Dischinger, Franz, 144, 162; Zeiss Planetarium, Jena, 144, *144*
Dischinger and Finsterwalder, 164
Duchamp, Marcel, 14, 15, 19, 22, 25; *Étant donnés: 1ᵉ la chute d'eau, 2ᵉ le gaz d'éclairage*, 25, *25*
Dyckerhoff and Widmann, Wiesbaden, Germany, 162; German shell, *140*; Market Hall, Frankfurt, 162

E

Edwards and Hjorth, 74
Eiffel, Gustave, 23; Statue of Liberty, 23, *23*, 24, *24*, 25
Einstein, Herbert, 15
Eliot, T. S., 22
Esquillian, Nicholas, 170
Euler, Leonard, 124

F

Faber, Colin, 14, 141
Finsterwalder, Ulrich, 142, 162
Flügge, Wilhelm, 162
Foster, Norman, 77; Hong Kong and Shanghai Bank, 77
Fuller, Buckminster, Geodesic Dome, 144

G

Garlock, Maria, 15
Gaudí, Antonio, 10, 16, 63, 94, 167, 169
Gauss, Carl Friedrich, 124
Gehry, Frank, 17, 18; DG Bank Building, Berlin, 18, 144, *145*
Giacometti, Alberto, 19; *Spoon Woman*, 19, *19*
Giedion, Sigfried, 166
Goff, Bruce, 81
Goldsmith, Myron, 19
Graham, Bruce, 19
Gropius, Walter, 14
Guastavino, Rafael, 16

H

Harrison, Wallace, 19
Harrison, Abramovitz and Abbe, 74; US Steel Building, Pittsburgh, 74–76, *75*, *76*
Heikkinen and Komonen, 45; Finnish Embassy, Washington, D.C., 45
Hellmuth, Yamasaki & Leinweber, 163; Lambert Field Terminal Building, St. Louis, 163
Holgate, Alan, 17
Huizinga, Johan, 172

I

Isler, Heinz, 12, 15, 16, 18, 25, 87–101, 142, 161, 169–72; BP Service Station, Deitingen, Germany, 16, *17*; Eschmann Chemical Factory, Thun, Switzerland, 91; Heimberg swimming pool, Heimberg, Switzerland, 97;

Heimberg tennis center, Heimberg, Switzerland, 96, 97, *97*, 171, *171*; Kilcher Company, Recherswil, *99*; Leuzlinger Sons Company, *97*; open-air theater, Grötzingen, Germany, 93–94, *94*, *95*; Sicli Company Building, Geneva, 16–17, *17*, 97, *99*, *100*; Swimming pool, Brugg, Switzerland, *98*; Trösch Company, 170; Wyss Garden Center, Solothurn, Switzerland, *99*
Isozaki, Arata, 19; Centennial Hall, Nara, 19, 111, *111*; Sant Jordi Sports Palace, Barcelona, 19, 108–9, *109*
Ito, Toyo, 57, 62; Pavilion for Serpentine Gallery, London, 57–59, *58*

J

Jahn, Helmut, 19
Jensen and Frank, 70
Johnson, Philip, 19, 76; AT&T Headquarters, New York, 76, *76*

K

Kahn, Fazlur, 19
Kahn, Louis, 19
Kalinka, John E., 162
Kaufmann, Erich, 88; Teepott restaurant, Rostock, 88
Kawaguchi, Mamoru, 15, 19, 25, 103–21; Centennial Hall, Nara, 19, 111, *111*; Expo '70, Osaka, 19, 104, 115, *115*; Electric Power Pavilion Annex, 116–17, *117*; Fuji Group Pavilion, 19, 116, *116*; Grand Roof of the Festival Plaza, Expo '70, 104, *105*, *106*; Flying carp, Kazo, Japan, 120–21, *120*, *121*; Inachus Bridge, Beppu, Japan, *102*, 103–4; Namihaya Dome, Osaka, *110*, 110–11; Pantadome system, 106–8; Pavilion for Twelfth World Orchid Conference, Tokyo, *119*, 119–20; Sant Jordi Sports Palace, Barcelona, 19; World Memorial Hall, Kobe, 108; Yoyogi Indoor Stadium, Tokyo, 19, 105, 112–13, *112–14*
Kluver, Billy, 19
Kohn Pederson Fox: World Financial Center, Shanghai, 84, *84*, *85*
Kommendant, August, 19
Koolhaas, Rem, 21, 22, 53–54; Casa da Musica, Porto, Portugal, 55, *55*; CCTV project, Beijing (under construction), 22, 55, *56*; Grand Palais, Lille, 54; House, Bordeaux, 22, 53, *53*; Kunsthal, Rotterdam, 54; Seattle Public Library, 54, *55*

L

Lacroix, Roger, 137
Lardy, Pierre, 18, 169
Larrambebere, Gonzalo, 45
Le Corbusier, 41, 53, 126; Maisons Jaoul, Neuilly-sur-Seine, 41; Ronchamp, 53
LeMessurier, William, 19
Leonhardt und Andrä, 17
Lerner, Ralph, 15
Libeskind, Daniel, 21, 22, 59; Victoria and Albert Museum extension, London, 22, 59, 60, *60*, 61, *61*
Liebniz, Gottfried Wilhelm, 124
Longfellow, Henry Wadsworth, 24

M

Maillart, Robert, 12, 13, 18, 129, 164, 170; Salginatobel Bridge, Schiers, Switzerland, 13, 128–29, *129*; Schwandbach Bridge, Hinterfultigen, 13, *13*
Mallarmé, Stéphane, 22, 23
Mandelbrot, Benoit, 57
Melan, Ernst, 162
Melan, Josef, 162
Mengeringhausen, Max, 144
Menn, Christian, 12, 15, 17, 18, 19, 25, 123–39; Averserrhein Bridge, Cresta, 129–30, *130*; Averserrhein Bridge, Cröt, *130*, 131; Biaschina Bridge, Giornico, 132, *133*; Chandoline Bridge, Sion, 132, *133*; Charles River Bridge, Boston, 127, *134*, 135–36; Felsenau Bridge, Bern, 19, 132, *133*; Ganter Bridge, 19, 132, *133*; Rhine Bridge, Reichenau, 131, *131*; Salginatobel Bridge, Schiers, 128–29, *129*; Salvanei Bridge, 18; Sunniberg Bridge, Klosters, 19, 127, *134*, 135
Mies van der Rohe, Ludwig, 21, 53; Seagram Building, 21, *21*
Miró, Joan, 167, 169
Morisseau, André, 88; Royan Market, 88
Morrow, Irving F., 128
de Moura, Eduardo Souto, 22; Braga Stadium, 22
Munk, Knud, 50; Carlsberg Brewery, Northampton, UK, 50, *50*
Murata, Yutaka, 116, 118, 119; Expo '70, Osaka: Electric Power Pavilion Annex,

INDEX | 185

116–17, *117*; Fuji Group Pavilion, 116, *116*; Pavilion for Twelfth World Orchid Conference, Tokyo, *119*, 119–20
Müther, Ulrich, 88; Teepott Restaurant, Rostock, 88

N
Navier, Louis, 124
Nervi, Pier Luigi, 12, 13, 15, 53, 88, 89, 161, 164–66, 171; George Washington Bridge Bus Station, New York, 165; Little Sports Palace, Rome, 165, 166, *166*; St. Louis Abbey, 88; St. Mary's Cathedral, San Francisco, 12, *12*; UNESCO, Paris, 165
Newton, Isaac, 124
Nordenson, Guy: World Trade Center Towers A and B, 20

O
Ortega y Gasset, José, 167
Otto, Frei, 17, 144; Stadium roof, Munich Olympics '72, 150, *150*

P
Paz, Octavio, 9, 14–15
Pei, I. M., 20, 77, 78, 81, 83, 84, 163; Bank of China Tower, Hong Kong, 77–81, *78*; Miho Museum Bridge, *83*, 83–84
Pei, Cobb, Freed & Partners, 81; Meyerson Symphony Hall, Dallas, 81–82, *82*
Peiri, Antonio, 10
Pevsner, Nikolaus, 39
Piano, Renzo, 18
Picasso, Pablo, 15, 167, 169
Picon, Antoine, 15
Piñero, Emilio, 103
Pound, Ezra, 22

R
Rice, Peter, 11, 18, 105; MOMI tent, London, 11; Centre Georges Pompidou, Paris, 18, *18*, 105; TGV Station roof, Lille, 11
Riley, Terence, 15
Roberts and Schaefer Company, Chicago, 162; Century of Progress World's Fair, 162; Hayden Planetarium, New York, 162
Robertson, Leslie, 15, 19, 20, 22, 25, 67–85; AT&T Headquarters, New York, 76, 76–77, 84; Bank of China Tower, Hong Kong, 20–21, 77–81; *77*, *78*, *79*, *80*, *81*; IBM Building, Pittsburgh, 67, 67–68, *68*; Miho Museum Bridge, Shiga-Raki, Japan, 20, *83*, 83–84; Myerson Symphony Hall, Dallas, 81–82, *82*; US Steel Building, Pittsburgh, 74–76, *75*, *76*; World Financial Center, Shanghai, *84*, 84–85, *85*; World Trade Center, New York, 20, 68–73, *69*, *70*–*73*, 75
Roebling, John: Brooklyn Bridge, New York, 23, 25
Rogers, Richard, 18
Rüsch, Hubert, 162

S
Saarinen, Eero, 19, 88; TWA Terminal, John F. Kennedy Airport, New York, 88
Safdie, Moshe, 19
Salvadori, Mario, 9
Sarger, René: Royan Market, 88
Sasaki, Mutsuro, 15
Schlaich, Jörg, 15, 17, 18, 19, 25, 141–59; Berlin-Spandau railway station platform roof, 149, *149*; Bullfight arena, Zaragoza, 152, *154*, *155*; DG Bank Building, Berlin, 144, *145*; Glass-fiber-reinforced concrete shell, Stuttgart, 142, *143*; GRC Shell for the Stuttgart Federal Garden, 18; Historical Museum, Hamburg, 148, *148*; Indoor pool, Nestor Hotel, Neckarsulm, Germany, 146, *147*; Roof over ice-skating rink, Munich, 150, *151*; Ice-skating rink, Hamburg, 152, *153*; Olympic Stadium, Munich, 17, 18; Pool, Kuala Lumpur, 152; Railway station, Ulm, 149, *149*; Stadium, Riyadh, 152, *152*; Stadium roof, Stuttgart, 152, *153*; Ting Kau Bridge, Hong Kong, 19; Zoo, Berlin, roof, 149, *149*
Schober, Hans, 18
Schreier, Aaron: World Trade Center, New York, 73
Sekler, Eduard F., 23–24
Sentieri, Pedro, 83
Severud, Fred, 19; Seagram Building, 21, *21*
Simon, Louis, 88; Royan Market, 88
Siza, Alvaro, 21, 22, 62; Portuguese Pavilion, Hannover, Germany, 62, *62*
Stirling, James, 51–53; Neue Staatsgalerie, Stuttgart, *51*, 51–52; No. 1

Poultry, City of London, 53; Polytechnic, Singapore, 53; Science Library, Irvine, California, 53
Strauss, Joseph, 128; Golden Gate Bridge, San Francisco, *122*, 125, 128
Stubbins, Hugh, 19

T

Tange, Kenzo, 19, 104, 108; Grand Roof of the Festival Plaza, Expo '70, 104, *105*; National Stadium, Singapore, 108; Yoyogi Indoor Stadium, Tokyo, 105, 112–13, *112–14*
Tedesko, Anton, 15, 88, 142, 161–64, 166, 171; Airport, St. Louis, 88; Brook Hall Farm dairy building for Century of Progress World's Fair, Chicago, 162; Hayden Planetarium, New York, 162; Ice-hockey arena, Hershey Chocolate Company, Hershey, Pa., *160*, 163; Lambert Field Terminal Building, St. Louis, 163
Torres-Garcia, Joaquin, 47
Torroja, Eduardo, 12, 88, 167, 169 170; Algeciras Market, 88; Coal Silo, Instituto de Ciencias de la Construccion, Madrid, 12, *12*; Zarzuela Hippodrome, Madrid, 88
Tsuboi, Yoshikatsu, 19; Yoyogi Indoor Stadium, Tokyo, 19

U

Utzon, Jorn and Hau, Todd, and Littleton: Sydney Opera House, Sydney, *48*, 50

V

Valery, Paul, 22
Vitruvius, 123, 124; *Ten Books of Architecture*, 123

W

Wachsmann, Konrad, 144
Webb, Craig, 18
Weidlinger, Paul, 19; Beinecke Rare Book Library, New Haven, 21, *21*
Williams, William Carlos, 13
Wirz, P.: Kilcher Company, Recherswil, 99
Wittgenstein, Ludwig, 24

Y

Yamasaki, Minoru, 19, 68, 73, 75, 88, 163; Airport, St. Louis, 88; Lambert Field Terminal Building, St. Louis, 163; World Trade Center, New York, 68–73, *69*

Trustees of The Museum of Modern Art

David Rockefeller*
Honorary Chairman

Ronald S. Lauder
Honorary Chairman

Robert B. Menschel
Chairman Emeritus

Agnes Gund
President Emerita

Donald B. Marron
President Emeritus

Jerry I. Speyer
Chairman

Marie-Josée Kravis
President

Sid R. Bass
Leon D. Black
Kathleen Fuld
Mimi Haas
Richard E. Salomon
Vice Chairmen

Glenn D. Lowry
Director

Richard E. Salomon
Treasurer

James Gara
Assistant Treasurer

Patty Lipshutz
Secretary

Wallis Annenberg
Celeste Bartos*
Eli Broad
Clarissa Alcock Bronfman
Donald L. Bryant, Jr.
Thomas S. Carroll*
David M. Childs
Patricia Phelps de Cisneros
Mrs. Jan Cowles**
Douglas S. Cramer
Lewis B. Cullman**
H.R.H. Duke Franz of Bavaria**
Gianluigi Gabetti*
Howard Gardner
Maurice R. Greenberg**
Vartan Gregorian
Alexandra A. Herzan
Marlene Hess
Barbara Jakobson
Werner H. Kramarsky*
June Noble Larkin*
Thomas H. Lee
Michael Lynne
Wynton Marsalis**
Harvey S. Shipley Miller
Philip S. Niarchos
James G. Niven
Peter Norton
Maja Oeri
Richard E. Oldenburg**
Michael S. Ovitz
Richard D. Parsons
Peter G. Peterson*
Mrs. Milton Petrie**
Gifford Phillips*
Emily Rauh Pulitzer
David Rockefeller, Jr.
Sharon Percy Rockefeller
Lord Rogers of Riverside**
Anna Marie Shapiro
Anna Deavere Smith
Ileana Sonnabend**
Emily Spiegel**

Joanne M. Stern*
Mrs. Donald B. Straus*
Yoshio Taniguchi**
David Teiger**
Eugene V. Thaw**
Jeanne C. Thayer*
Joan Tisch*
Edgar Wachenheim III
Thomas W. Weisel
Gary Winnick

*Life Trustee **Honorary Trustee

Ex Officio

Peter Norton
Chairman of the Board of P.S.1

Michael R. Bloomberg
Mayor of the City of New York

William C. Thompson, Jr.
Comptroller of the City of New York

Christine C. Quinn
Speaker of the Council of the City of New York

Jo Carole Lauder
President of The International Council

Franny Heller Zorn and
William S. Susman
Co-Chairmen of The Contemporary Arts Council